THE ORANGUTANS

THE ORANGUTANS

Gisela Kaplan
Lesley J. Rogers

PERSEUS PUBLISHING
Cambridge, Massachusetts

Library of Congress Catalog Card Number: 00-102285
ISBN 0–7382-0290-8

Perseus Publishing is a member of the Perseus Books Group.

Perseus Publishing books are available at special discounts for bulk purchases in the U.S. by corporations, institutions, and other organizations. For more information, please contact the Special Markets Department at HarperCollins Publishers, 10 East 53rd Street, New York, NY 10022, or call 1–212–207–7528.

Set in 10.5-point Goudy by the Perseus Books Group

First printing, May 2000
1 2 3 4 5 6 7 8 9 10—03 02 01 00

Find us on the World Wide Web at http://www.perseuspublishing.com

CONTENTS

ACKNOWLEDGMENTS

The inspiration for this book arose from our experiences with orangutans in Sabah, East Malaysia. This book is not a report on our research findings as such, but without that personal contact with orangutans it could never have been written. We therefore gratefully acknowledge research funding from the University of New England and the Australian Research Council (ARC) that enabled us to conduct our research in Borneo. We are also grateful to Dr. Anne Russon for sharing with us some of her own experiences with orangutans and permitting us to include them in the book. In addition to our field research, we also conducted studies on the orangutans at Perth Zoo, and we are indebted to Dr. Rosemary Markham, Leif Cocks, and the director of the primate section, Reg Gates, for supporting that research. We also thank Craig Lawlor of the University of New England for assistance with some of the figures and Associate Professor Ian Metcalfe for providing the basis for some maps. All the photographs were taken by Gisela Kaplan. Finally, we wish to acknowledge the very good collaboration with the staff of Allen & Unwin in the production of this book; we thank Ian Bowring in particular, and Colette Vella.

Part I

EVOLUTION

1
ARRIVING AT AN ISLAND HOME

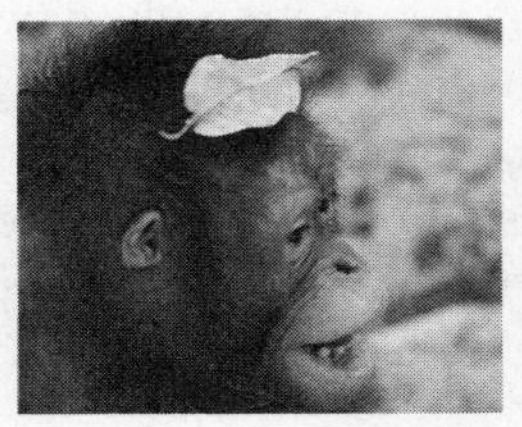

On a narrow road winding through the lower mountains and rain forest of Kalimantan, an orangutan exacts a toll from vehicle drivers who enter his territory. Without fail, he emerges from the thick jungle to stand in the middle of the track with palm outstretched. He is a lone and awesome figure, conspicuous now, but at other times completely hidden from view in the dense green foliage. A small piece of food or a trinket will satisfy him, and he will then disappear until the next intruder is heard in the distance. He is a wild orangutan and seeks no other human contact.

He symbolizes the pressing need to protect his habitat. The rain forests of Southeast Asia are the richest and most diverse in the world, teeming with the earth's greatest variety of plant and animal species. Wild orangutans are usually found in the uppermost treetops. They mingle with many other animals in the vaulted canopy high above ground level, feeding on leaves and fruits. Birds have no trouble flying from treetop to treetop, and some of the mammals have developed ways of gliding from tree to tree using sails made of skin flaps stretched between hindlimbs and forelimbs. The gliding squirrel moves about in this way, and there are even flying frogs and gliding snakes among these high-rise dwellers.

Orangutans are the heaviest tree-living animals, and they move from tree to tree by swinging on the branches. Often they climb clasping with hands and feet in all directions or they hang beneath the branches and move about by grasping with one hand after the other (known as *brachiating*). Sometimes they propel themselves from one tree to another by standing more or less upright on a supple trunk or branch and setting it in motion like a swing. In this way they can make a large arching movement. After swaying back and forth several times, they reach out for a branch of another tree while letting go of the first one. Only rarely has an orangutan been seen to leap from one branch to another, as do so many smaller primates that live in the trees.[1] Watching from below, we notice how agile they are.

Island Life

Orangutans live in the forests on just two islands of the Malay Archipelago—Sumatra and Borneo (Figure 1.1).[2] Kalimantan is the Indonesian part of Borneo, and Sarawak and Sabah are the Malaysian parts of the same island. On these islands orangutans survive in small areas of rain forest that have escaped the bushfires and the logger's ax. In Sumatra, their range is restricted to the Gunung Leuser National Park and a few areas adjoining it, all in the north of the island. In Borneo, orangutans are more widely spread, but even so, their ever-diminishing range covers much less than one-third of the island.

As a result of the felling of the rain forests to make way for agriculture and to obtain timber for building, furniture, and even for disposable wooden crates, the number of orangutans is declining rapidly. The orangutan is now on the list of endangered species. Estimates indicate that the total population of orangutans in Sumatra and Borneo together declined from 80,000 in the 1980s to only 20,000 in the mid-1990s. Since the devastating bushfires of recent years, their number is probably much less than 20,000.

Today orangutans survive in the last outposts of their range, formerly much more extensive. They used to occur on mainland Asia, throughout all of Southeast Asia, including Myanmar, Laos, Thailand, Cambodia, and Peninsular Malaysia, as well as in Java and in their present homes in Sumatra and Kalimantan. Fossils of orangutans have been found in all these regions. Their range also extended into the southern

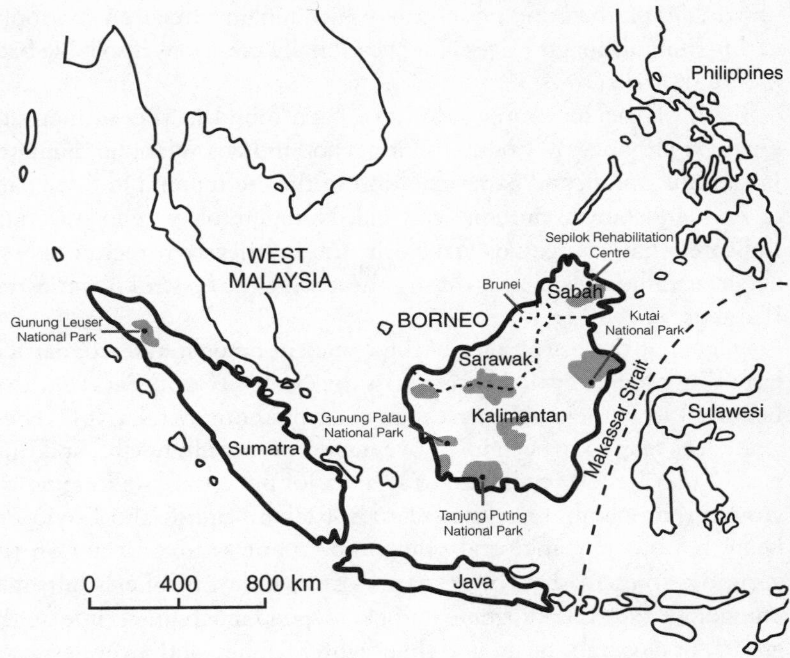

FIGURE 1.1 *Southeast Asia showing the Malay Archipelago. The gray areas designate where orangutans are present today. Dashed line indicates the Wallace Line.*

part of China as far north as Beijing, and even east to regions that are now the islands of Taiwan and Hainan. This was their distribution 2–3 million years ago, and it seems to have remained so until about 10,000 years ago. Only since that time have they become entirely island dwellers.

The First Orangutans

We can find out something about the first orangutans by unearthing their fossils and examining them in detail. Usually only fragments of fossils are found, and the most common fragments are teeth. These are made of hard materials that survive the passage of time, and orangutans' teeth are protected by an exceptionally thick layer of enamel. A tooth can be recognized as belonging to an orangutan by the presence of wrinkles on the grinding surface of each molar, a likely adaptation to eating fruit, although leaves and occasionally meat may also have been part of their diet.

Teeth of ancient orangutans have been found in the Yunnan and Guangxi provinces of Southern China and in Laos, Vietnam, Sumatra, Java, Kalimantan, and Sarawak.[3] Most of the teeth found in these parts of Asia are about 2 million years old. None are older than this. This indicates that orangutans arrived in the southeastern region of Asia about 2 million years ago, during the geological epoch known as the Pleistocene.

Judging by the size of some of these ancient orangutan teeth, particularly those found by T. Harrisson in the Niah caves in Sarawak, they belonged to heftier specimens than their present-day relatives.[4] These giant apes may have been up to two meters or more in height, and their weight may have forced them to spend a lot more time walking on the ground than climbing in the treetops. Richard Smith and David Pilbeam reason that ancestral orangutans spent as much time on the ground as present-day chimpanzees.[5] Even today, heavier adult male orangutans (90–100 kilograms) probably spend much more time on the ground or closer to the ground than lighter females and juveniles.[6]

From the eighteenth century on, there have been occasional reports of very large modern-day orangutans walking upright on the ground. The well-known orangutan researcher John MacKinnon was shocked when he almost bumped into a huge orangutan walking along a path.[7] Likewise, Biruté Galdikas, who has spent most of her life studying orangutans, spoke of her encounter with an extremely large walking orangutan. He was ambling along a path, looking down, at first unaware of her presence. On catching sight of Biruté, some four meters away, he stared at her and then suddenly whirled around and was gone.

But where did the large ancient orangutans come from? We must go back to the very first apes. Fossil records suggest that the first apes appeared 17–23 million years ago, during the early geological epoch known as the Miocene. The tail is the most easily distinguishable feature separating apes and monkeys. Apes do not have tails, whereas monkeys do. Apart from humans, the apes alive today are the gorillas and chimpanzees, living in Africa, and the gibbons and orangutans, living in Asia. All these existing apes evolved from more ancient ancestral forms.

The first primates without tails evolved in the warm rain forests of East Africa. The fossil of an ancestor of modern apes was found there, and it is called *Proconsul*. From its anatomy, *Proconsul* appears to have moved about mainly by walking on the ground using its feet and hands in quadrupedal style, although it may also have climbed trees. When *Proconsul* walked on the ground, it did so using the palms of its hands to give support, a method used by orangutans today. The weight is carried on the outside edges of the palms. This means of supporting the weight while walking contrasts with that used by gorillas and chimpanzees: They curl their fingers toward the palm and support their weight on their knuckles, not their palms.

About 17 million years ago, ancient apes like *Proconsul* gave rise to a branch of evolution that eventually led to the gibbons of today. This happened well before the ancestor of the orangutan appeared. Before the ancestral orangutan evolved, apes like *Proconsul* migrated from Africa into Europe. By dating the fossils of apes found in Europe, we know that this took place 14–16 million years ago. Somehow these ancient apes managed to cross the stretches of water that separated the European continent from North Africa (Figure 1.2). They may have reached Europe via the Gibraltar-to-Spain route or via the Middle East route. Although these are the shortest crossings, neither of these routes to the European continent would have been particularly easy. In fact, at that time, the continent of Africa was further from Europe and the Middle East than it is today. There were, however, times when the sea levels were lower and land bridges may have existed. Perhaps the ancestral apes crossed from Africa to Europe on these land bridges or perhaps they drifted on floating rafts of debris. Rafts could have formed from the dense plant growth occurring in the forests along the coastline of North Africa at that time. These are only ideas—no one knows how the apes managed to get to Europe.

In Europe, the ancestral apes evolved into many different forms and spread across the continent. Some of them migrated to Asia and may have gotten as far as Southeast China 5–10 million years ago. The first creatures that could be the ancestors of modern orangutans evolved a long way from their present home in Southeast Asia, in a region known as the Siwaliks, now in North Pakistan. Fragments of fossil skulls of apes called *Sivapithecus* have been found in this region. There were several species of *Sivapithecus*, which is the name of their common genus.

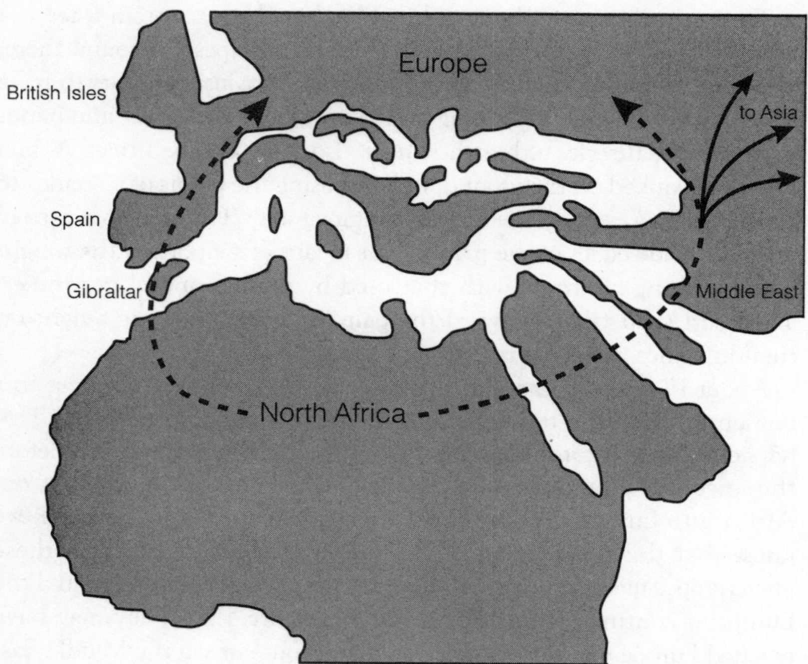

FIGURE 1.2 *Location of landmasses in the African and European regions some 14–16 million years ago. The arrows represent possible routes by which the ancient apes might have migrated from Northern Africa to Europe and on to Asia.*

Anthropologists have painstakingly pieced together *Sivapithecus* skulls to discover their similarities to, and differences from, present-day orangutans. Bone material taken from skull specimens of *Sivapithecus* has been used to determine the age of the skull, revealing that *Sivapithecus* lived 8–12 million years ago, in the late Miocene epoch.

The earliest ancestors of the orangutan looked rather like present-day orangutans, although the skull was slightly broader and the limbs less mobile. They were larger than their modern descendants, although some of their relatives became even larger still, as the now-extinct Chinese orangutans show. The large size and reduced mobility of the limbs of *Sivapithecus* indicate that these apes spent more time on the ground than their lighter present-day relatives. Nevertheless, they still depended on the forest to survive. *Sivapithecus* met its ultimate fate and became extinct when the dense forests and woodlands of northern Pakistan and India disappeared. This came about as the world's climate became colder, drier, and more seasonal. Grasslands, deciduous forests, and scrub began to replace evergreen forests. By the end of the Miocene, all the ancient forms of apes in Europe and most parts of Asia were extinct, apart from some known to have been still surviving in China and a few other regions of East Asia. Monkeys in the same regions survived the colder drier climates because they had adapted to life in more open habitats. Apes require more specialized climatic conditions than monkeys, and so they survived only where the climate remained wetter and warmer than elsewhere.

Some people believe in the existence of the Yeti of the Himalayas and say that it might be a special, shy kind of orangutan that has adapted to living at high altitudes in freezing temperatures. This is almost certainly fantasy. As far as we know, orangutans need warm rain forests to survive. Today they do not even occur at the higher altitudes in the mountains of Borneo.

Fossil remains of orangutans that must have looked almost identical to present-day orangutans are about 2 million years old. Most scientists agree on this date for the first appearance of modern orangutans, but there are some who would like to set the date earlier, at 5 million years ago. This difference in timing is really not very large on an evolutionary timescale—whether 2 million or 5 million years ago, the date of the first appearance of orangutans that looked the same as we know them today was quite recent in geological time.

Reaching the Islands

It would be interesting to know when orangutans spread from mainland Southeast Malaysia to Borneo, Sumatra, and Java. Looking at a map of the present time, it seems they had to cross vast stretches of ocean between the mainland and these islands, Borneo in particular. But at the time the orangutans made their way to these regions, Southeast Asia looked very different from its appearance today. Two million years ago, at the very most 5 million years ago, the landmasses that were to become the islands of Borneo and Sumatra were still connected to mainland Asia, as shown in Figure 1.3. There was a large bridge of land

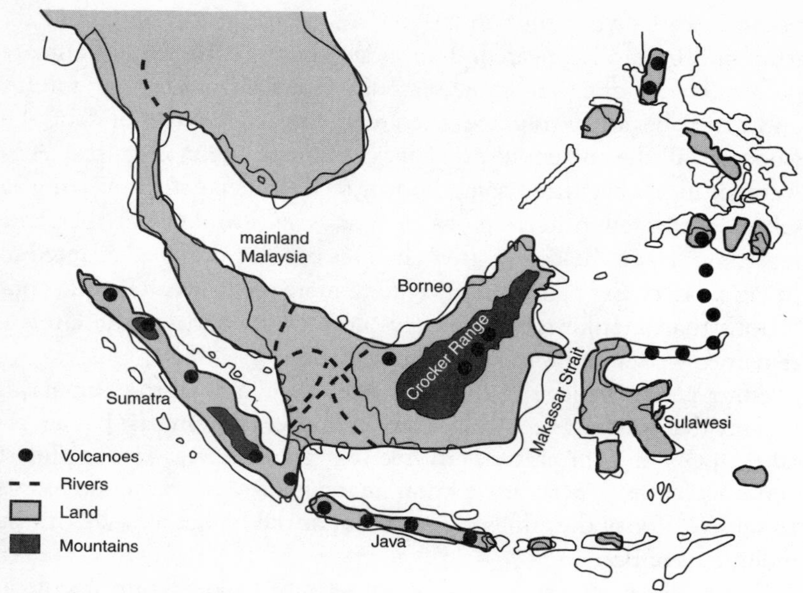

FIGURE 1.3 *The landmasses of the Malay Archipelago some 5 million years ago (indicated by thick lines and gray shading) superimposed on those of the present day (thin lines and no shading).* SOURCE: *Based on Hall 1997; Morley and Flenley in Whitmore 1987:55.*

stretching from mainland Malaysia to Borneo and another, smaller bridge across to the southern end of Sumatra.

Many maps of the land formations at that time show Java connected to the same landmass as Sumatra and Borneo in the Pleistocene epoch,[8] but Robert Hall, of the South East Asia Research Group at London University thinks differently. According to Hall's recent geological information, the northern tip of Java was separated from the southern tip of Sumatra by a narrow strait of sea, just as it is today.[9] Further geological research is needed before we can be sure of Java's position at the time orangutans reached it and left their fossils there. The matter is complicated by fluctuating sea levels caused by the freezing and melting of the polar ice caps (see below).

The early orangutans, together with many other species, could have migrated from the Asian mainland across the land bridges and into Borneo and Sumatra. The concept of migration, as it is used here, has nothing to do with migration in birds or humans. It refers to the colonization of new areas over a long period of time and many generations of orangutans—a gradual spreading of the population.

If there was no land bridge between Sumatra and Java at the time, perhaps the early orangutans crossed the short stretch of sea between the two islands by floating on branches or other debris. It probably happened accidentally, possibly by the orangutans being swept away in violent storms. Most people would assume that orangutans do not construct rafts to cross water, after deciding that they want to get to the other side—we have reserved such ways of thinking for humans. We assume humans are the only species that can plan ahead and wonder what is around the corner or on the other side of the ocean or river. But Biruté Galdikas has seen an orangutan at her rehabilitation center dragging logs to a stream and then placing them to make a bridge to allow him to cross to the other side. That orangutan had certainly formed the intention of crossing to the other side.

These observations make us wonder whether the ancient orangutans may have intended to cross the ocean between Sumatra and Java and even that they might have constructed some sort of raft to do so. Most anthropologists believe that orangutans learn such skills only when they have been in close contact with humans; they merely imitate the use of boats by the people living and working in their vicinity. Perhaps no wild orangutan would construct a raft or plan to cross a stretch of water, but we would like to keep the possibility open.

Of course, there may have been a time around 2 million years ago when the stretch of ocean between Sumatra and Java was narrower than it is today, and so crossing might not have been as daunting a task as we think. There were times when the levels of the ocean were much lower than today and more land was exposed; glacial periods dominated the earth's climate, locking up the water of the oceans in the polar ice caps and lowering the level of the seas as a result. It is even possible that for brief periods of geological time, the sea level was low enough to afford the orangutans a land crossing between Sumatra and Java or between Borneo and Java. Rafts or floating debris may not have been needed.

The glacial periods were broken by times when the polar ice caps melted and the sea levels rose again. Geologists refer to these times as interglacial periods. The islands of Borneo, Sumatra, Java, and the thousands of smaller islands in the Malay Archipelago were formed during an interglacial period, and that is the state of the world today. In fact, at the present time the sea level is about as high as it has ever been and only about half of Southeast Asia is exposed. The last high point of a glacial period (i.e., a glacial maximum) occurred 21,000 years ago.[10]

The polar ice caps have formed and melted about twenty times over the last 2 million years, and the sea levels have fluctuated by more than 200 meters.[11] As a result, islands in the Malay Archipelago have come and gone. But the glacial periods were longer than the melting periods, and they provided the opportunity for orangutans and other animals to migrate from mainland Asia by a land route to Borneo, Sumatra, and maybe even Java.

Orangutans did not spread any further east than Borneo to the island of Sulawesi, probably because Sulawesi was separated from Borneo by a large stretch of very deep water, known as the Makassar Strait. A line can be drawn along the Makassar Strait separating the flora and fauna of Asia from the flora and fauna of the islands to the east of it (see Figure 1.1). This line is called the Wallace Line.[12] To one side of it are the Asiatic plants and animals of mainland Asia, Borneo, Sumatra, Java, and Bali and to the other side of it are entirely different species on the islands of Sulawesi, Irian Jaya, Australia, and many smaller islands. Orangutans are on the Asian side of the line.

The presence of land bridges between the mainland and the islands of Borneo and Sumatra might explain how orangutans eventually reached their island homes, but their passage from the mainland would not have been without enormous obstacles. In the Pleistocene era, when this migration occurred, major rivers flowed across the land bridges, especially across the bridge between the mainland and Borneo. The huge North Sunda River flowed northward. Other rivers flowed across the trail between Sumatra and Borneo. They were all mighty, fast-flowing rivers, and migrating animals would have had to cross them. Somehow the orangutans did it, either by chance or design. If they had not done so, they would have become extinct. Only by reaching the areas that were to become their island homes did the orangutans survive.

The type of plant growth on the land bridges would also have created difficulties for the orangutans. The land bridges were unlikely to have supported the orangutans' preferred habitat of tropical rain forest because the rainfall would have been too low and seasonal during the glacial periods. The bridges are thought to have been corridors of savanna, covered in grasses and scrubby trees. This would have been hostile to orangutans because their diet probably consisted mainly of the leaves and fruits of the forest. On the other hand, orangutans today eat hundreds of different types of food, and their dietary adaptability may, perhaps, have allowed them to find sufficient food in the savanna.

The ancient orangutans may have had to climb trees to escape predators, particularly during the night—if that is why they sleep in nests above ground level. We should not forget that leopards and tigers were migrating over the same territories as the orangutans. Perhaps the migrating orangutans were able to find pockets of lowland forest on the savanna corridors and move between them. In fact, judging by the numbers of present-day orangutans living in different types of forest, it seems that they prefer lowland forest to higher mountain forest.[13] Most of them live in the lowland forests, which have tall dipterocarp trees (endemic to Southeast Asia), at altitudes of less than 150 meters. These days, the logging and clearing of the lowland forests is driving them into the higher areas, but that is not where they choose to be. They did once live on the higher slopes of Mount Kinabalu, a mountain more than 4,000 meters high in Sabah, but they are no longer found at altitudes greater than 1,500 meters.

Seas and rivers were not the only obstacles for the migrating orangutans. High mountain ranges were another barrier. Orangutans would have met a major mountain range on the landmass that became Borneo. Today it is called the Crocker Range and runs through the middle of Borneo roughly parallel to the western coastline. Volcanoes within this range were active several million years ago. No orangutan would have braved the high altitudes and precipitous barriers.

Barriers and Genetic Variations

The presence of the high Crocker Range may explain the genetic differences that exist between at least two or three separate populations of orangutans in Borneo today (see Chapter 2). Genetic differences refer to differences in the inherited material (genes) inside the cells of the body. Genes are passed on from one generation to the next, carrying the information that influences the way individuals develop and function. Genetic variations, for example, make orangutans different from humans, and they are also the cause of differences between orangutans in different areas. Genetic differences begin to develop when populations of the same species are separated for long periods of time. If the two populations are separated for an extremely long time, they may become so different that they eventually form separate species. Of course, experience also causes differences to develop in different populations, and this needs to be taken into account too.

Until quite recently, scientists thought that all the Bornean orangutans belonged to the same population and that the main difference existed between the Sumatran and Bornean orangutans. But then Collin Groves, of the Australian National University, and colleagues measured the skulls of orangutans from the eastern and southwestern regions of Borneo and found a number of differences.[14] It seems that the Crocker Range has separated eastern from western, or southwestern, populations long enough for physical differences to emerge. But the separation has been only long enough for some genetic differences to occur, not long enough for separate species to form (see Chapter 2). The matter is still under debate.

The orangutans in Sumatra and Borneo have, it seems, been separated for a longer time, long enough for the population of orangutans in Sumatra to be genetically quite different from both Bornean populations.[15] In some ways, this can be seen in variations in appearance. The Bornean orangutans have flatter, broader faces than the Sumatran orangutans, and the latter are slightly smaller and tend to have moustaches. Differences are also seen in the size and shape of their skulls. The Bornean orangutans have been given the scientific name *Pongo pygmaeus pygmaeus*, and the Sumatran orangutans *Pongo pygmaeus abelii*. *Pongo* is the *genus*, which is a broader category than *species*. The species comes next—*pygmaeus* (note that both Sumatran and Bornean populations are *pygmaeus*). The final term is *pygmaeus* again in the case of the Bornean orangutans and *abelii* for the Sumatran orangutans. These terms refer to *subspecies*, indicating that differences exist between the two populations.

It is not known what isolated the Sumatran and Bornean populations of orangutans from each other. Devastating droughts may have been an important factor. Droughts occurring during the coldest glacial periods may have had a major effect on the distribution of orangutans and other species in this region of the world. Some scientists think that droughts had a more important influence on the evolution and distribution of the primates of the region than mountains or any other physical barrier. It is said that a cold drought may have eliminated the original orangutans in Sumatra. Then Sumatra may have been recolonized by orangutans from Borneo at a later date.

Back to the Trees

We have mentioned that the ancestral orangutans were larger than present-day orangutans and may have spent much of their time on the ground. Then later circumstances may have driven the orangutans to return to spending most of their lives in the trees. Some researchers believe that predators at ground level may have encouraged a relocation to the upper canopy. To make living in trees possible, the orangutans had to become smaller and, in fact, they became just small enough to be able to do this. Orangutans today are the largest arboreal primates; they may be close to the maximum weight that allows them to stay up in the forest canopy without falling.

Exactly when orangutans returned to the trees is unknown. Some people argue that humans are the only ground-dwelling creatures that present a serious threat to orangutans, and humans may have been instrumental in forcing the orangutans back to the trees. We think that their return to high-rise living must have happened earlier and agree that it may have been in response to being hunted by clouded leopards and the Sumatran tiger. In the past, these predators were present in sufficient numbers to be a serious threat to orangutans, although that is no longer the case, especially in Borneo. The fact that Sumatran orangutans, to this day, spend more of their time in the trees than the Bornean population supports this explanation because only in Sumatra does there remain a formidable ground predator, the Sumatran tiger. Others have suggested that the orangutans returned to the trees to escape ground-dwelling parasites, such as leeches.

How Many Orangutans Are There?

Until 10,000 years ago, the distribution of orangutans differed little from that occurring in the early Pleistocene. Only after that time did it shrink dramatically. Today their range is far less extensive and under further threat from the felling of rain forests for timber and the expansion of agriculture. Estimates of the population size of orangutans, made in 1993 by the Orangutan Population and Habitat Viability Analysis Workshop, arrived at figures of 9,000 orangutans in Sumatra and between 10,000 and 15,000 in Borneo.[16] The total area of habitat occupied by orangutans in Borneo was estimated to be just over 22,000 square kilometers, subdivided into about eight separate areas referred to as "islands" of forest.

Unfortunately, these figures are not very reliable because it is difficult to assess numbers of living orangutans accurately, particularly in inaccessible areas. One way of estimating population size is to fly over the forests along straight lines in planned directions (known as *transect lines*) and to estimate the number of orangutans from the number of nests spotted. The accuracy of this method varies according to the habitat being surveyed. Nests are more easily seen in open forest than in dense forest. There are also difficulties in areas with cliffs.

Added to this are the problems created by the nest-building behavior of the orangutans. Sometimes they build a new nest every day; at other times they use the same nest more than once. We have even seen individuals build more than one nest a day, one for a siesta and one to sleep in at night. These factors make accurate estimates of the population size impossible. A new technique is being trialed now. It uses a camera that senses thermal wavelengths and can be used to see warm-blooded animals in the dark. It is hoped that it can "pick up" orangutans and distinguish them from other creatures. If that is possible, the camera can be attached to an airship piloted remotely and flown over the forests.

The density of orangutan populations varies in different types of forest. It is highest in one region of Sumatra where there is an abundance of fig trees, on which orangutans gather to feed. As we shall see, these regional differences in population density may have led to variations in social and other behavior. They may be the reason for at least some of the behavioral differences now being found between Sumatran and Bornean orangutans.

2

APES ON THE EVOLUTIONARY TREE

To look into the eyes of an orangutan is to come face-to-face with one's own humanity. The species differences are blurred. Something intangible—a feeling of knowing and yet not knowing—passes between human and red ape. The fleeting exchange leaves a sense of wonder.

We saw our first totally wild orangutan one evening at dusk in the rain forest, upstream on a narrow tributary of the Kinabatangan River. We were sitting quietly in a small dinghy when we heard the cracking of branches. Only a very large animal could make that sound. Our eyes strained against the darkening foliage of the dense forest. Then we saw him, high above us, building a nest for the night. Our first glimpse of him was a huge hand rising above the bowl of the nest and then crashing down on the branches to secure them into place. The hand appeared several times as we held our breath. Then his head appeared over the brim of the nest, and for what seemed like a very long time, he peered down at us. We were chilled with excitement. He was an enormous adult male, large cheek pouches very clear, eyes just visible in the fading light. At a giddying height in that impenetrable forest, there was a creature very like us, looking down at us, just as we were looking up at him. We were under no illusion that it was a friendly meeting; we were the intruders, and we felt unwelcome. He resumed his nest building. We were never to forget this first encounter.

A privileged experience like this leads one to feel that humans and other apes have much in common, but what does science tell us about the differences and similarities? There are several ways in which the topic can be approached using scientific methods. One is by comparing the skulls, teeth, and other bones of ancient and present-day apes and humans and so construct a picture of their evolution (see Chapter 1). Another way of looking at the differences and similarities is to use the new methods of molecular genetics to compare the molecular structures of the genes and other parts of the cells. We discuss that here. A third approach (see Chapter 3) is to look at the similarities and differences in the structure of the brains of the living apes and compare their abilities to solve problems, to learn, and to carry out other complex behaviors.

First, let us take a look at the evolutionary tree.

The Evolutionary Tree

The evolutionary tree is a way of representing how different species are related to one another and how they have evolved. The trunk of the tree is pictured as the main line of evolution to humans, *Homo sapiens*, perched at the very top of the tree (Figure 2.1). The branches represent divergent lines of evolution from the main trunk; they mark the various species that branched off along their separate paths of evolution. The points where the branches leave the trunk give an idea of when each species appeared. Species that branch off lower down the tree appeared early in evolutionary time, and those that branch off higher up appeared later and are more closely related to humans. This is the kind of evolutionary tree that most people are used to. It puts humans in the superior position above all other species. If an orangutan were to draw an evolutionary tree, it might put itself at the top and us further down. There are many different ways of drawing the arrangement of the tree's branches.

Figure 2.2 is a more accurate representation of the evolutionary tree for the hominoids (apes). It is more like a branching vine than a tree—it indicates when groups of species separated from each other, and all the species still alive today are listed across the top. The ancestral and

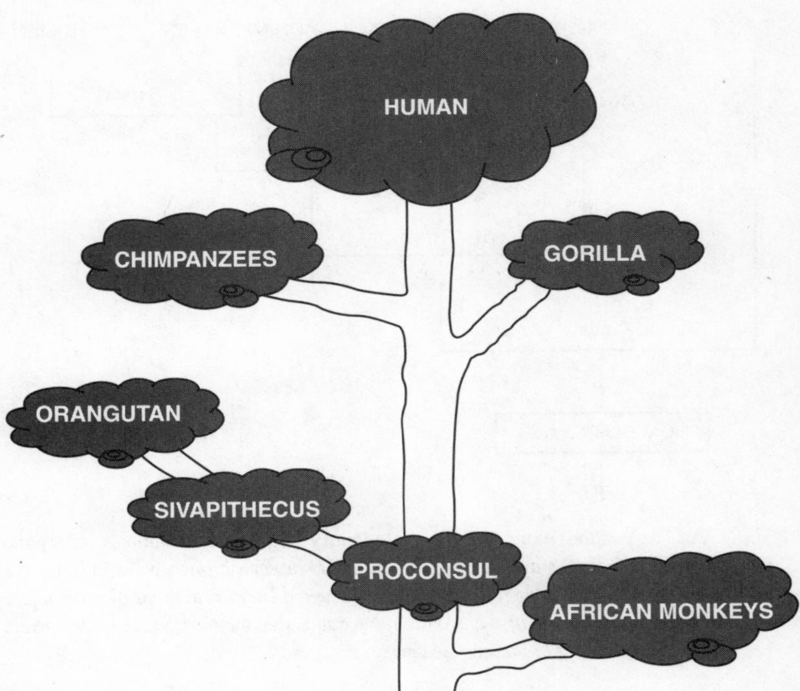

FIGURE 2.1 *Simplified evolutionary tree, as it is often imagined to be. There are other much more accurate representations of evolution.*

now-extinct forms have not been indicated in the figure. The numbers at the branch points indicate when two lines of evolution separated. There are several lines of evolution. The one that concerns us most here is the hominoid line, which includes all the apes. As we saw in Chapter 1, *Proconsul*, now extinct, was the first type of ape. The gibbons branched off from the hominoid line about 17 million years ago. Orangutans branched off about 8–12million years ago, probably closer to 12 million. Then comes the branch that led to modern-day gorillas, diverging about 8 million years ago, and lastly the branches that diverge to chimpanzees (both the common and pygmy chimpanzees) and humans at about 5–6 million years ago.

We construct such an evolutionary tree by discovering fossils, or fragments of them, and establishing the age of the fossils. The dating of fos-

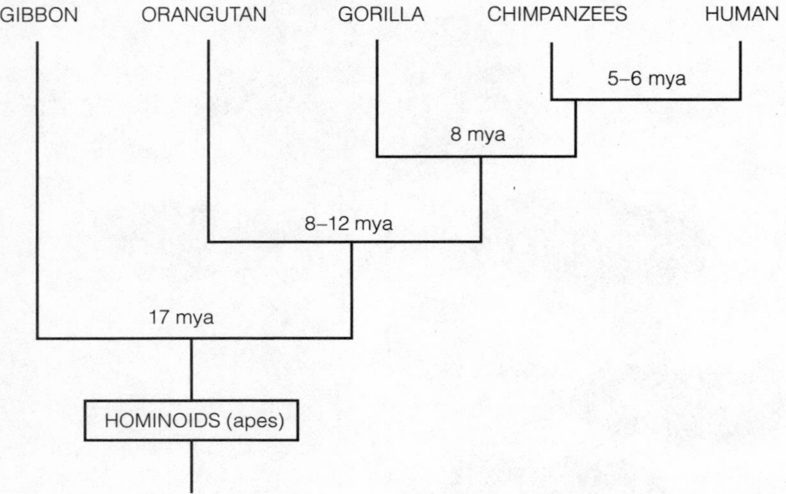

FIGURE 2.2 *A representation of the evolutionary tree for the hominoids. Only the living apes are shown. The numbers at each branch point indicate the likely time when the divergence occurred (mya = millions of years ago). This version of the evolutionary tree is the one most likely to be correct according to present evidence.*

sils helps to decide where to place the branches on the evolutionary tree (see Chapter 1). But today there are other ways of determining relationships among species and pinpointing how many million years ago they branched off on their separate evolutionary paths. In the field of molecular genetics, using sophisticated biological techniques, it is possible to work with tissues taken from living animals and find out how the different species evolved.

Comparing the Genes

Techniques in molecular genetics allow us to compare the genetic material of living species and determine how well the genetic material of one species matches that of another species. This tells us how closely related the two species are. Genetic material is passed from one generation to the next, and as time passes, slight changes creep into its structure. These changes are called *mutations*. If the two species being compared separated on their own branches of evolution a long time ago (i.e., they branched apart near the base of the evolutionary tree), they will have accumulated many mutations that do not match up. If there is close matching of the genetic material from the two species, there must have been less time for the accumulation of differing mutations and we can say that the two species are quite closely related. In other words, they must have branched off from each other recently.

This technique of matching the genetic material from different species is called DNA hybridization. DNA (deoxyribonucleic acid) is the name of the genetic material, or genes. The genes are strung together to make chains called *chromosomes*. They carry an inherited code, or program, that guides some aspects of the development of the individuals representing the species. Members of the same species share their genes (or DNA) when they breed with each other. Thus, even though genetic mutations occur, they are shared with other members of the same species. If two populations of (originally) the same species are separated in some way and cannot breed with each other, each will accumulate different genetic mutations that are not shared through breeding. As time passes, the number of different mutations will increase and the similarity of the genes (or DNA) between the two populations will decrease. Eventually, the two populations become so different from each other that they are regarded as two different species.

We can think of it in a different way. The DNA is made up of a string of subunits, each like a word in a sentence. Strung together, these "words" spell out the genetic messages that are passed on from generation to generation. The more messages that match up between two species, the closer the genetic relationship between the two species. The mutations accumulating in the DNA with the passage of time can be thought of as changes in the "words" or "word" sequences. The degree of matching between the DNA sequences, or sentences, from each species can then be translated into evolutionary distance or proximity. Genetic difference indicates separation in evolutionary time. The further back in time two species separated onto different branches of the evolutionary tree, the greater the difference in their DNA (the fewer matching sentences). This information can be used to draw up an evolutionary tree.[1]

The DNA used in hybridization tests is obtained from inside the nucleus of each cell. This DNA material is inherited from both the mother and father of each individual (via the egg and the sperm) and is called nuclear DNA, or genomic DNA. In addition, there is another source of DNA. It is present in the material surrounding the nucleus (known as the cytoplasm), and there it is found inside organelles (small structures) called mitochondria. This kind of DNA is called mitochondrial DNA and is inherited only from the individual's mother (via the egg only). Mutations accumulate faster in the mitochondrial DNA than in the nuclear DNA, and this fact allows scientists to make comparisons that determine when different species branched apart. Because the mitochondrial DNA mutates at a fast rate, it can also be used to look at variations between different populations within the same species (e.g., to look at the differences between Bornean and Sumatran orangutans and the different populations of orangutans within Borneo).

Getting hold of tissue for DNA analysis is no problem. Some researchers prefer a sample of skin, and this is often obtained from wild animals by using darts. A small hollow dart (called a plug biopsy dart) is fired at the target animal from a distance and at a speed that prevents it from penetrating too far into the animal's body. The dart soon falls off and can be collected from the ground, where it is easily identified by its brightly colored tail. The tissue sample inside the hollow dart can then be taken out and used in laboratory tests.

Only a very small sample of DNA is needed because it can be copied in the test tube, a process called *cloning*. In this way, the small sample of DNA can be multiplied to give enough DNA for hybridization tests. Even a strand of hair (with the hair follicle attached) can be used, and hairs have been collected from orangutans' nests—after the orangutan has woken up and left, of course. But even then, it is not easy to collect a hair sample from a nest because it means climbing many meters up into the trees. A fecal sample dropped to the ground is collected more easily and will suffice. All fecal samples contain cells from the animal's body, and these can supply enough DNA to be multiplied and used in DNA hybridization tests.[2]

The DNA hybridization approach has shown that 98 percent of the genetic material of humans is the same as that of orangutans. Put another way, human and orangutan genes differ from each other by only 2 percent. There is an even greater match between the DNA of humans and chimpanzees. Here the difference is only 1 percent.[3] This puts chimpanzees closer to us than orangutans, and it fits the evidence from fossils (see Chapter 1). It means that humans and orangutans branched off from each other earlier than did humans and chimpanzees (see Figure 2.2). Gorillas lie in between these two points of separation. Their DNA differs from human DNA by 1–2 percent.

The DNA hybridization technique is revolutionary. It has made an enormous difference to the study of evolution, although it has not solved all controversies. For example, some molecular geneticists say that orangutans diverged from the hominoid line (i.e., the line leading to humans) as recently as 8–9 million years ago.[4] Compare this with the 12 million years obtained from the fossil record.[5] Others have suggested that the divergence occurred much earlier in evolutionary time, at 19–24 million years ago.[6]

Each estimate based on DNA hybridization depends on the accumulation of mutations but also on setting the "molecular clock." Before we can determine the time based on the number of mutations that have accumulated in the genetic material, we have to know what the starting point was—in scientific parlance, we have to calibrate the clock. This is done by using independent evidence obtained from fossil records to select a time at which a branch occurred in the evolutionary tree. It should be a time that is known with as much certainty as possible (e.g., when whales diverged from other mammals, or New World primates from Old World primates). This time is used as a reference point to start the molecular clock, and this is where the first problems arise. It depends what divergence is used to set the clock.[7] Clocks set at different times lead to different estimations of the time when orangutans first evolved.

There are other problems with estimates of where to put the branches on the evolutionary tree. One is that mutations may occur at different rates in different populations (or in different species). This means that the clock may tick over faster in some places on the evolutionary tree than it does in others.[8] Another problem is that some DNA from the father may get mixed up with the mitochondrial DNA, which comes mainly from the mother.[9] This would look like a mutation, although it is not. This all goes to show that exciting as the new techniques in molecular genetics are, they have not solved all the questions that are of interest to us. Nevertheless, we are optimistic that answers will be found in the near future.

In focusing attention on genetic material, we should not lose sight of the fact that the development of individuals (or of a species) is influenced by the environment as much as by the program, or code, carried in the genes. Experience and learning are important influences on development. In fact, the process of development involves continuous interaction between the genetic program and influences from the environment. These ever-changing interactions determine the physical form, the physiology, and the behavior of an organism.

Not all the genetic code carried by an individual is expressed (i.e., read out), and experience can induce some parts of the code to be read out and suppress other parts. The genetic material examined in the test tube by the DNA hybridization technique is the entire genetic code, the complete genetic potential of the species or individual. Knowing this code gives us a guide to evolution (see above), but it cannot tell us which parts of the code will be expressed in the living animal or in what sequence they will be read off during development. It is impossible to tell this from DNA material in test tubes.

Anthropologists place much importance on the times when the different species of apes evolved. This is the main reason that researchers are more interested in the entire DNA code than in development and the particular set of genes that is actually read out. Generally, they pay much more attention to chimpanzees, and even gorillas, than to orangutans because they are more interested in the species that are closest to humans in terms of evolution (determined from the whole genetic code). Because orangutans branched off from the hominoid line on the evolutionary tree earlier than gorillas and chimpanzees, they are usually regarded as a less interesting comparison with humans. As a result, orangutans have been "the neglected apes," at least until recently.[10]

People tend to think that living chimpanzees are the apes most like humans because they separated from our branch of evolution only 5–6 million years ago. This approach ignores the fact that not all the genetic code is expressed in any one species or individual. It does not take into account all the complex influences that determine which parts of the genetic code will be expressed. In other words, it focuses on evolution and pays no attention to development and the importance of influences from the environment.

Even though there is a greater difference between the genetic codes of orangutans and humans than between chimpanzees and humans, it is not impossible that this relationship would be reversed were we to consider only those parts of the code that are actually read out (i.e., if we took into account the influences of experience, learning, and other effects of the environment). More of the same parts of the code may be expressed in orangutans and humans than in chimpanzees and humans. If this is the case, living orangutans may be more like humans than living chimpanzees are, at least in certain ways. So far, this is only speculation, but it could explain some of the observations that we discuss below.

Looking Alike

In the 1980s, Jeffrey Schwartz noticed that orangutans and humans have a number of similar characteristics that are not present in chimpanzees or gorillas.[11] These characteristics include the presence of a particular vein in the arm, certain structures of the teeth and bones, a long period of pregnancy, the absence of genital swelling during periods of sexual receptivity (compared with the large genital swelling of chimpanzees), and taking part in sexual intercourse at any time during the menstrual cycle. There is also a characteristic that chimpanzees and gorillas have in common that is not present in either orangutans or humans—"knuckle walking." When apes walk on the ground, they use their arms to support the weight of their bodies, and this requires them to put their hands on the ground. Chimpanzees and gorillas do this by balling their hands and resting their weight on the knuckles (i.e., knuckle walking), as mentioned in Chapter 1. They do not touch the ground with the palm of the hand. All the weight of the upper part of the body is on the knuckles. Orangutans, on the other hand, do touch the ground with the palms of their hands. Although they may bend the fingers so that their knuckles also touch the ground, their weight is supported on the outside edge of the palm. In this sense, too, orangutans are similar to humans. The structure of the human hand shows no indication of knuckle walking; when human children crawl, they place the palms of their hands on the floor.

In fact, the structure of the bones in the human hand gives no hint of any knuckle walking in our ancestry. This is difficult to explain from the evolutionary tree shown in Figure 2.2. Since chimpanzees and gorillas diverged from the hominoid line on separate branches, at separate times, and both knuckle walk, we might expect knuckle walking to have appeared before they branched off. This would mean that the ancestors of humans were knuckle walkers. If so, humans must have lost that characteristic very early on, although even then, we would expect traces of it to remain somewhere in the structure of the bones of the hand. Another explanation might be that gorillas and chimpanzees both evolved knuckle walking independently after they had diverged from the hominoid line. This would mean that the common ancestor of gorillas, chimpanzees, and humans did not knuckle walk. But it seems rather unlikely that knuckle walking would appear in two separate lines of evolution.

Weiss suggested a way around this problem by devising a different arrangement of the branches on the evolutionary tree. As shown in Figure 2.3, he places humans on a branch before the chimpanzees and gorillas.[12] The last branching then occurs between gorillas and chimpanzees, instead of between humans and chimpanzees. This alternative tree fits the DNA hybridization results as well as our first evolutionary tree. Despite this, a more recent molecular technique, looking at a new part of the genetic code, has produced results in favor of the gorilla line branching (or separating) off before humans and chimpanzees branched apart, as in Figure 2.2.[13] There is no consensus on the correct version of the evolutionary tree, but the weight of evidence points in favor of Figure 2.2.[14]

All molecular techniques depend on computer programs to analyze the complex results obtained, and not all programs give the same result. But the majority of different ways of looking at this problem point to the chimpanzee as the closest relative to humans, then the gorilla and next the orangutan.[15]

Nevertheless, many puzzles remain and they show that there are many ways of interpreting the scientific results. Whatever way we look at it, orangutans seem to pose more questions than answers about the apes and their origins. This is one of the reasons they are so fascinating to us.

Differences Among Orangutans

It has been known for a long time that the orangutans of Sumatra and Borneo are different from each other. In Chapter 1, we mentioned how measurement of the skulls of orangutans led one researcher to suggest there might be greater differences between the populations of orang-

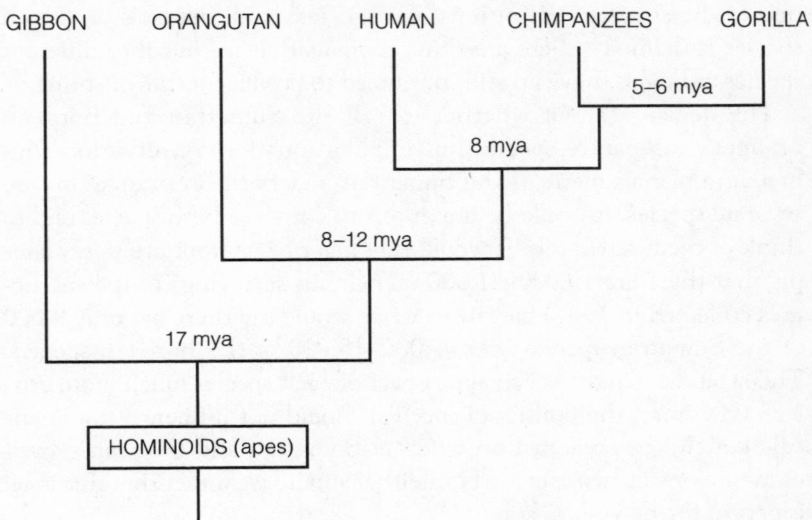

FIGURE 2.3 *An alternative version of the evolutionary tree, proposed by Weiss (1987). Although this might explain the similarities between orangutans and humans, it is less likely to be correct than the one shown in Figure 2.2.*

utans in different parts of Borneo than between those in Sumatra and Borneo. Comparison of the mitochondrial DNA by the hybridization technique is an ideal way of testing this suggestion because, as we have said before, mitochondrial DNA accumulates mutations faster than nuclear DNA. Therefore, differences emerge sooner than in nuclear DNA, hence mitochondrial DNA is better for comparing populations that may have been separated for relatively short periods of time.

Lu Zhi and colleagues compared mitochondrial DNA samples from orangutans at nine locations in Borneo and two locations in Sumatra.[16] Contrary to the suggestion above, their results showed greater differences between the Sumatran and Bornean orangutans than between the different populations in Borneo. Zhi and colleagues even felt that the Sumatran and Bornean orangutans were different enough to be different species. Some other researchers agree,[17] but not everyone is of the same opinion.[18] We know that Sumatran and Bornean orangutans can interbreed, because they do so in zoos. This means they are not quite different enough to be regarded as different species, if we go by one of the standard definitions of a species. It all depends on how a species is defined—there are other animals that we put into different species although they can still interbreed to produce fertile offspring.[19]

The decision about whether to call the Sumatran and Bornean orangutans separate species has implications for conservation zoo-breeding management. If the Sumatran and Bornean orangutans are separate species, it would be important to conserve both species and to think of them separately. It would no longer be appropriate to say simply that there are 19,000–21,000 orangutans surviving (to use the figures collected in 1993) but, instead, we would say there are only 9,000 of the Sumatran species and 10,000–15,000 of the Bornean species. This would make the survival prospect of each species much more critical. Of course, the politics of survival should not influence the scientific conclusions reached on whether Sumatran and Bornean orangutans are one or two species, but neither should we forget that this issue is part of the debate.

As to variation between populations within Borneo, we believe that more sampling and DNA hybridization studies are needed before we will know anything for certain. Zhi and colleagues were unimpressed by differences between the populations of orangutans that they sampled in Borneo, but their results do indicate that the population in Sabah differs from that in Sarawak and in the southwest of the island. We would like to see more studies conducted, but we are also aware that orangutans are moved from one locality to another when their forests are destroyed by logging and fires. This will confound future research of this kind.

So far we have talked about differences between populations of orangutans, but of course, within each population of orangutans, no two individuals are exactly the same. This is why scientists need to take samples of hair, skin, or fecal material from as many individuals as possible in order to make comparisons between populations. They need to know the number of differences within the population as well as between populations. Individual orangutans differ in physique, physiology, and behavior because each has its own unique genetic material and experiences. Some of the genetic material and some of the experiences are common to all orangutans, but each individual has, overall, his or her own particular combination of influences on development. Each orangutan is an individual in the way he or she looks and behaves.

Orangutans are as different from one another as humans are. After a short time with the orangutans, we were able to recognize at least thirty individuals accurately and another twenty more with slightly less certainty. It was important for us to learn to do this as quickly as possible; our research depended on being able to score the behavior of individuals, and that meant we had to recognize each one from day to day. We also had to recognize them when we returned to the same research site each year. After a year's absence, most adults looked the same, but as one would expect, the infants and juveniles had not only grown but had changed in appearance. We had to take a refresher course from the people working at the research station before we could recognize them again with certainty, although there were always a few orangutans with whom we had a special affinity and could recognize anywhere.

BJ was one of these (Plate I). He was an exceptionally mischievous orangutan, always lying in wait to grab a camera or tripod. Or he would swing from the liana vines, making sure that he was in full view and ready to receive attention. He was an ideal subject for photography. A few years ago, friends gave us a calendar featuring apes and on the front cover was a photo of BJ. We recognized him immediately, just as we would a member of our own family. The same is true of others, such as Rampig, who also liked our cameras and tripod (Figure 2.4).

Mostly we recognize individual orangutans by their face, just as we do humans. Every face is different (Plate II gives some examples). But faces cannot always be seen from the ground when the orangutan is high up in the forest canopy. Then we rely on the distinguishing features of body size and shape, the presence or absence of cheek pouches, skin color, hair color and length, and body posture and movement. These all vary from orangutan to orangutan. Age and sex, of course, contribute to these differences. Adult males are much larger than adult females, and only the males have large and visible cheek pouches.

Among the orangutans we studied, the pouches can be seen on Simbo, an adult male aged close to thirty years (see Figure 5.1). You can see how impressive the cheek pouches make the face appear. When viewed face-to-face, Simbo looks huge, which must be useful when two males meet and eye each other off. Biruté Galdikas describes seeing two males interrupt their wrestling over a female to stare each other down like Sumo wrestlers about to attack.[20] Size counts in such encounters, and a large fearsome face can endow a psychological advantage. Looking as big as

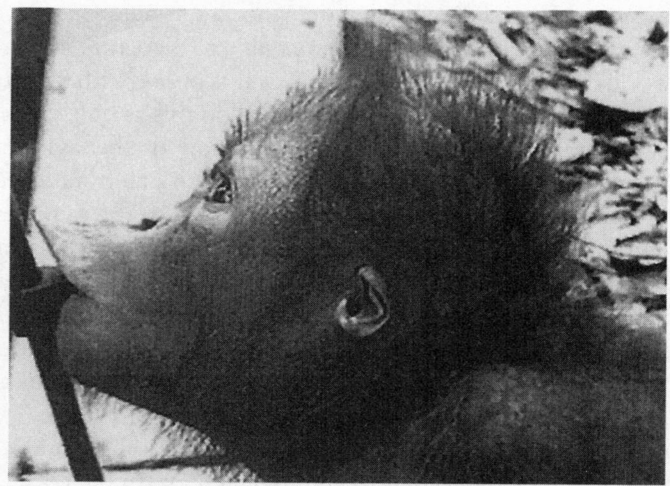

FIGURE 2.4 *Rampig tries to chew a tripod, something she knows she is not supposed to do. Her expression indicates that she is expecting an adverse response.*

possible is a tactic used by many species when they want to issue a threat, as we describe in one of our earlier books.[21] It is also seen in humans—when men are about to fight, they make fists of their hands, puff up their chests, and stretch out their shoulders and upper arms.

Cheek pouches vary in size and structure and so can be used to recognize individuals. The pouches become fatter when the orangutan is overweight, as so often seen in orangutans confined in zoos. Cheek pouches begin to grow as males become sexually mature. We saw them growing on nine-year-old Boy at the same time that he began to court a female. The pouches grow slowly and may continue to develop after sexual maturity is reached.

Skin color of the face and body varies also. Some orangutans, like Judy, have pink skin all over the face and body (Plate III). Raja has black skin (Plate II; also Figure 8.1). Jessica's skin has a reddish tinge (Plate V), and BJ has different skin colors on different parts of his body. Others have pink-and-black mottled skin, particularly on the belly. Skin color of the eyelids is also very important. Most infant orangutans have light pink to almost white eyelids, and so do some adults. This contrasts with darker skin on the rest of the face and makes shutting the eyes very obvious. It is used in communication (see Chapter 7).

Hair color does not differ as much as skin color, but the amount of hair and its length is a good indicator of individual differences. Some orangutans have very little hair. Others have thick hair that grows down their backs. Coiffure varies, too. Clementine has a fringe and straight bangs. Kurashi wears his swept forward (Plate II), and Abbie prefers the more natural look.

All these characteristics, and many more, make each orangutan different and special. We should not forget this when we study their evolution by mixing up tissues in test tubes or when we talk about "the orangutan" in a collective sense or "orangutans" as if they were all the same.

Who Is in the Human Family?

So far we have talked about animals as members of different species. This is one way of classifying them. Species are grouped into broader categories called *genera* (sing. genus), and these, in turn, are grouped into *orders* and then *families*. We humans have a family to ourselves, called Hominidae. Chimpanzees, gorillas, and orangutans are in the family Pongidae. Based on the results of DNA hybridization, some people have suggested that chimpanzees and gorillas should be included in the Hominidae family, leaving only orangutans in Pongidae.[22] This view recognizes the close similarity of the genetic code in humans, chimpanzees, and gorillas, and we are in favor of that. On the other hand, we find it rather difficult to leave orangutans out of this broader family, particularly as they have many features in common with humans that chimpanzees and gorillas do not.

If the family Hominidae were expanded to include the other great apes, a number of important issues would arise. We would have reason to extend to the other apes some, if not all, of the rights that we presently afford humans. This would mean that we could no longer keep them in captivity, in zoos, or as pets, and we could not kill them. We might also consider giving them rights to their habitat, although this is something we often do not acknowledge even for our fellow humans. The Great Ape Project is in support of giving rights to all the great apes.[23]

3
ADAPTATIONS OF THE ORANGUTAN

Orangutans are tree apes. Their bodies are adapted for tree living and so is much of their behavior. Nevertheless, living in trees may not be something that comes naturally to orangutans. They may have to learn to climb and move around in that dangerously high environment. Infant orangutans separated from their mothers and raised by humans in rehabilitation centers must be taught to climb. In fact, they find climbing so frightening that they have to be encouraged to do it, and once up a tree, they quickly come down again, time and time again. Living in the rain forest is not always a possibility for orangutans that have been reared by humans. We learned this from Judy, an orangutan at Sepilok Orangutan Rehabilitation Centre in Sabah.

Judy came to the sanctuary of the rehabilitation center after being rescued from a life in captivity with humans. When we met her, she was five years old (Plate III). She was particularly lacking in hair, which made her look more like a naked human than did her fellow orangutans. It comes as something of a shock to see a member of a species that can look so humanlike also behave like a human. For instance, Judy often walked upright, with feet slightly splayed and her tiny buttocks pinched in. Her entire body posture, her arm movements, and the way she turned her head and looked at things at times made it difficult to think of her as an orangutan. But her behavior is perhaps not all that surprising. Judy had lived all her early life with humans. We do not know when she was caught from the wild, but since she was only five years old when she joined the rehabilitation center, it is a fair guess that most of her crucial childhood years were spent in the care of humans. Humans had been her teachers and her carers and she had learned her lessons well, perhaps too well. It is impossible to tell whether Judy had made up her own mind to walk upright like humans or whether she had been encouraged to do so.

Judy did not appear to us or to the rehabilitation staff to be a good prospect for rehabilitation back to the wild. Added to this, she seemed to prefer to be around humans rather than among orangutans or, even more naturally, on her own. At times she would play with sticks and other objects next to humans, always looking to see whether they were watching, just as children do.

Despite her strong attachment to humans, she had to go through the rehabilitation program at the center. This meant that after a period of adjustment to the forest around the research station, she had to be taken on a long walk deep into the center of the reserve and left there. Most orangutans who have had previous experience of living in the wild will be in no hurry to return to the station. On first release, some make their way back to the research station, taking several days to do so. In time, however, they return less and less often and, finally, do not come back.

It was different for Judy. When she was taken out into the forest, she merely gave the carer a head start and turned up at the station a little after he had returned. The procedure was repeated several times, but Judy chose not to get the message. She made her way back to human company each time and apparently was distressed by the experience of being left behind. We know of four occasions of her failed release. Imagine a juvenile or teenage orangutan being unhappy at being left in the forest of her birth and origin! And note that Judy did not get back to the station by swinging from branch to branch or walking on all fours through the jungle. No: Judy walked upright on the paved path through the forest to the research station. "Ladylike" this may have been, but it was a rather poor performance for an orangutan.

On the fourth and last attempt by the staff at the research station, Judy came up with a novel idea. She decided to take a shortcut back. From the release point, she found a back route out of the forest to a roadside and a bus stop. Taking a seat at the bus stop, she waited alone until the local bus appeared. It was on its way to the research station and had a number of passengers, although the front seats were empty. The bus stopped and the driver opened the door. The bus driver was kind. He had no particular prejudice against orangutans and understood her intention to board the bus. He did not insist on the normal bus fare. So Judy did not have the long walk back. She got on the bus and without the slightest hesitation or sign of insecurity took the seat behind the driver. There she sat calmly as the bus made its way to the research station. When it arrived and the door was opened, she alighted, heading straight for the research station. The staff said that she expressed much pleasure at meeting her human friends again.

After this episode, it was accepted that Judy had failed all attempts at rehabilitation to the wild. She was granted her wish—to live around humans, interacting for a brief period each day with the influx of tourists and delighting them with her antics, and sharing quiet moments sitting alongside her favorite people working at the research station.

In a way, she had chosen the less difficult life. Orangutans that are released into the forest face a very uncertain future. How long they survive alone is unknown. It seems that some thrive while others do not. We humans may romanticize the beauty of the rain forest and see it as the Eden to which all orangutans must return. Judy does not share this view. To her, the forest is a lonely place where the food she likes is difficult, if not impossible, to find. Climbing and moving around in the trees is very difficult. There are all manner of biting insects and strange creatures that are not friendly. This is no place for an orangutan that has known the comforts of home. It is doubtful whether she could ever learn to like her ancestral home as a place to live, not just to visit.

Other captive orangutans raised in the wild may hanker for a life in the rain forest. Locked in small cages, they may dream of tall trees to climb, liana vines to swing on, and food specialities to find. It is a matter of how one learns to survive in certain environments, both physically and mentally. Most of an orangutan's desires are not innate but are learned during early life.

Tree Living

Wild orangutan infants learn to climb gradually during repeated excursions from their mother's body. These forays are at first only brief. The infants literally go out on a limb, at first maintaining some contact with the mother even if only via her outstretched leg, then alone to the end of a branch, and so on until independence is reached. Very early in life, of course, orangutan babies learn to live with heights. They have plenty of experience of looking down at the forest floor and lower canopy while riding on their mother's body.

The orangutan's body is well designed for climbing. Their arms are long relative to the rest of the body, their fingers are long, and every aspect of the hand is designed for gripping branches. Their feet can grip like hands, as we discovered the first time we set off into the reserve at Sepilok to photograph the orangutans. We marched enthusiastically into the reserve, cameras hanging exposed over our shoulders and tripods under our arms. Little did we suspect that BJ was lying in ambush, just 100 meters ahead. As we approached, he pretended to look elsewhere and then suddenly grabbed a leg of the tripod. Gisela held tightly to one end of the tripod and Lesley held onto BJ. We attempted to make him let go, but the moment the tripod was released from his hands, he grabbed it with his feet. This left Gisela holding one end of the tripod, BJ holding the other end, and Lesley holding BJ by the shoulders. We had to start again to get the feet to release their grip and then the hands took over. It was like dealing with an octopus. After the tripod, the cameras held most attraction for BJ, but he never lost interest in the tripod, and from then on, it was always necessary for Lesley to guard the tripod while Gisela took photographs. BJ became our constant companion.

Orangutans can use their feet to hold food while eating and to manipulate it. They have no trouble putting a foot in their mouth, even when hanging from a branch, and they use their feet to manipulate many things (Plate IV), including pieces of food. But mostly they use their hands for feeding, either one hand when hanging by the other or both hands after wedging the body into a fork in a tree. Usually, their feet are used for grasping branches, as an extra support in addition to the hands or to allow hanging upside down. Eating seems to be almost as comfortable upside down as the "right" way up. The hips of the orangutan are designed to allow full rotation of the joint and the legs can be moved at almost all angles, something we humans can do only with our arms. In addition, the knee and ankle joints allow greater movement than is possible in bipeds like us. The orangutan's toes are long and curved like their fingers, allowing them to clasp branches of many sizes.

With all these anatomical adaptations, orangutans are extremely versatile in their manner of moving through the trees. They are also quite agile on the ground despite having feet, and hands, that are better adapted for climbing than for ground walking.

Perhaps the most impressive aspect of the orangutan's hand is its ability to carry out fine manipulation as well as power gripping. Power gripping is necessary to support the body weight when hanging. It is surprising that a hand that is so strong can also manipulate things so delicately. We discovered this by accident on our first visit to Sepilok. We were on our way to a conference in Europe, so we had carried all our money, our passports, and other papers with us, as well as binoculars and cameras. A sign at the entrance to the reserve recommended that we leave valuables at the small building outside, but there were no lockers or any other safe place to leave our valuables. We decided to put everything in one zippered bag that Lesley would carry.

This bag became the focus of one delinquent orangutan's attention. His only achievement in rehabilitation had been to spend his days in the forest limbering up to tourists, arms and hands held above his head or used as walking crutches, making obscene sounds with his lips. Before the day was over, he had grabbed the bag. The zip was no trouble to open. Desperately, Lesley held on to the strap to prevent the bag being whisked high into the trees, but she was losing the battle when Gisela arrived. Determined not to let the orangutan get away with our only hope of boarding the next plane, Gisela took hold of the bag too and, at eye height to the orangutan, began to wrench his grip off the bag while glaring him in the eye. The orangutan had never met a person like this. He was shocked. He looked at Gisela and then at Lesley with an expression of amazement, as if to say "What kind of human is this?" Numbed by amazement, he let her remove each finger's grip until the bag was ours again. But in exchange, he took the notebook in which we had been recording preliminary observations. At the time, it seemed almost as valuable a possession as the passports, plane tickets, and money. The orangutan disappeared into the undergrowth with the notebook, and there we found him examining it by gently turning the pages, each one separately, with such delicate movements that he might have been examining the pages of a priceless first edition. The power grip and "hand of the artist" belonged to the same individual.

Meals and Medicines

Some animals have very limited choice in their diets, but this is not so for orangutans. They eat more than 400 different kinds of food. Many of these food items are plants of various kinds, but orangutans also eat insects, and on occasion, they even kill mammals to eat. They are not exclusively vegetarian, as once thought.

The manipulative abilities of their hands and fingers are very useful for feeding on some of their favorite foods. Wild durian fruit attracts orangutans when its pungent odor signals that it is ripening. Durians become ripe between August and December, with a peak time in October. On average, only one durian tree grows in each one-and-one-half square kilometers of forest,[1] so orangutans have to travel quite long distances to satisfy their palate and it is likely that some have to leave their home range (discussed in more detail below). It seems that the orangutans know when each fruiting tree will ripen, and they return regularly to the same tree at the right time. This would require them to have excellent memories for remembering where the trees are located *(spatial memory)* and when the fruit will ripen *(temporal memory)*. This is no small achievement in the densely wooded forest.

Durians are covered in a hard skin with large prickles. Orangutans open the fruit by gripping it firmly between the teeth and using one hand to manipulate it until the skin breaks open at a weak point. By applying this skill, orangutans can feed on the durians picked from the tree, or already fallen to the ground, before the fruit is ripe enough for the shell to crack open of its own accord.[2] This gives them first serve before elephants and other species, including insects, gather to take their share. Other species must wait until the durians break open naturally because they cannot open the fruit themselves. Orangutans avoid competition with other species by being able to get in first. They gorge themselves on the fruit, sometimes until there is none left for the later dinner guests.

Although orangutans may outwit other species at their favorite fruiting trees and so avoid competition, they do sometimes compete against each other to obtain the fruit. Sri Suci Utami and colleagues studied competition between Sumatran orangutans in fig trees and found that one individual would move away from a fruiting tree when another orangutan approached and that there seemed to be a sort of hierarchy in priority of access.[3] No overt aggression occurred. The displaced one would wait and return to gather a meal when its superior had eaten. For the same reason, females usually avoid a tree already occupied by an adult male.[4]

Fig trees with ripe fruit become gathering sites for orangutans. As with the durians, they seek out the fig trees, remembering not only when each will have ripening fruit but also where it is among all the trees of the forest. Although priority of access to fig trees is sometimes seen, the right cannot always be exercised because orangutans often gather in fig trees in groups (see below). Ripe fig trees then become the focal point of social interaction in otherwise quite solitary lives. When mothers with infants feed in the same tree, the infants can play with each other. Adults too may have social exchanges when feeding in the same tree.

Like us, orangutans have a predilection for sweet foods. Durians are sweet, although they are eaten before they are quite ripe. Several kinds of jackfruit are consumed when ripe. Orangutans will even brave bee stings to rob a hive of its honey.[5] One orangutan was seen covering its head with large leaves to ward off stings while raiding a hive.

More than half the orangutan's feeding time is spent eating fruit, as Peter Ungar of the University of Arkansas found in a study based at the Ketambe research station in Sumatra's Gunung Leuser National Park.[6] He also found that half of these fruits are unripe, as in the case of durians, and that the orangutans eat a lot of fruits other than figs, with both soft and hard skins. Eating leaves, he found, took up most of the rest of their feeding time, but orangutans do vary their diets by including insects, bark, and, less often, flowers and nuts. The plant foods that they choose to eat tend to be low in toxins that might cause poisoning or other damage to their bodily functions if consumed in large quantities.[7] Toxins are a real risk for animals that eat mainly plants, and toxins can occur in fruits as well as leaves. Orangutans avoid them by selecting those species with lesser amounts of toxins in them and by eating the younger leaves.

Medicinal plants may be eaten on occasions when they are needed. No details on this are available for orangutans, but it is known that chimpanzees eat plants that medicate against internal parasites and possibly even malaria.[8] Other plants selected by chimpanzees act as antibiotics, and they are ingested in higher quantities in areas where the incidence of bacterial infections is high.[9] The eating of soil may also be for the purpose of curing intestinal ailments, and this behavior has been observed in orangutans by ourselves and other researchers.[10] They also smear soil over their bodies, perhaps to rid their skin and fur of parasites. We have seen them covered with mud. BJ had a habit of smearing wet white clay over his face to make himself look handsome. At least, that was the impression he gave as he smiled, white-faced, head held high.

On rare occasions, orangutans have been seen eating the flesh of other mammals. Sri Suci Utami and Jan van Hooff of the University of Utrecht have observed several instances of wild Sumatran orangutans eating slow lorises.[11] Three females were observed in all. This is not sufficient to conclude that females are more likely to eat meat than males, but females do spend more time foraging for insects than males do.[12] It might be significant that two of the females had offspring at the time. Perhaps the physiological demands of providing milk for a suckling offspring cause a temporary change in diet. Some orangutans eat the placenta after giving birth, and this too might be a behavior triggered by the physiological demands of pregnancy and lactation.

As Utami and van Hooff describe, a wild orangutan, called Yet, was in a fruit tree when she caught sight of a slow loris right in front of her. She slapped the loris until it fell to the ground, and when it attempted to flee, she ran after it, making high-pitched hoots. On catching it again, she killed the loris and then ate it. Such events are rare but memorable to those who witness them. There is at least one other example of carnivorous behavior in orangutans. Sugardjito and Nurhuda describe a sighting of an adult female consuming an infant gibbon, but it is not known whether the gibbon was dead or alive before the orangutan obtained it.[13]

Meat eating in orangutans is not common and may occur only at certain times when there is a higher than usual demand for protein and a scarcity of other foods. It seems to result from taking advantage of the moment (i.e., when the prey happens to be close at hand) rather than involve active hunting. In this case, orangutans differ from chimpanzees, who have been seen to form bands to pursue and ambush their prey, usually colobus monkeys.[14]

Complex social communication is required to coordinate a hunt, as is great agility of movement. Orangutans are capable of agile movement, but their forms of social communication may not be of the type necessary for hunting in packs. The difference between the two species in terms of hunting might, therefore, come about because chimpanzees are generally more social than orangutans. Above all, orangutans' desire to eat meat, even on rare occasions, may be less than that of chimpanzees. This explanation is supported by the fact that there have been more observations of meat eating and killing by chimpanzees than by orangutans. The chimpanzees at Gombe National Park have been seen to kill and eat some 400 colobus monkeys over a ten-year period.[15] On the other hand, there have been far fewer studies of orangutans than chimpanzees in the wild, and this may be why the behavior has been seen less often in orangutans than in chimpanzees.

Solitary or Social?

It is often said that orangutans are solitary apes. Compared with other apes, that is true, but they are not hermits. Four different kinds of social groups have been identified in orangutans. The first is the mother with her offspring (see Chapter 4). The second is the consorting pair (i.e., male and female sexual partners that stay together for a brief period of time).[16] The third is a traveling band, a small group that moves around together eating at the same fruiting trees; and the fourth is a temporary grouping formed by numbers of orangutans aggregating at a fruiting tree but moving off separately to go their individual ways.[17]

Exactly what kind of social organization is seen in orangutans may depend on the kind of forest they live in and the availability of food. Carel van Schaik of Duke University believes that the image of orangutans as solitary apes has come from studies of those living in the upland and mountainous areas of Borneo. There, each adult male holds a large territory that does not overlap with the territory of another male but does overlap several smaller territories each held by a female.[18] Van Schaik has observed orangutans living in a swamp forest in Sumatra, and there they seem to be more social than in Borneo.[19] Up to ten adults may feed in the same tree in Sumatra, and social cohesion is strong enough for bands to travel together. The reason for this difference, van Schaik suggests, is dietary. In mountainous Borneo, orangutans may have to forage over wide areas to find leaves to eat. This form of feeding is best carried out alone. In contrast, the Sumatran orangutans may find more fruits, and insects in the swamp and there they gather into groups to share the feast.

All apes have a wide variety of social behaviors, and despite the fact that orangutans are more solitary than the other apes, they too have many different ways of relating socially. They can adapt their social behavior to different situations. Biruté Galdikas has described twenty-seven different kinds of social combinations in orangutans.[20]

As more and more information about wild orangutans is collected, it seems that Solly Zuckerman's 1932 description of their social behavior as dominant males defending territories for harems of females is not correct.[21] Perhaps that particular form of social organization does occur in some places, but orangutan social behavior varies from one time to the next and from one place to the next.

Some researchers believe that adult males with territories may defend them aggressively, whereas females do not. This extra aggression by males may explain why they are so much larger than females[22] and perhaps also why they have cheek pouches that make them look bigger when viewed from the front. Males have been observed to fight each other (see Chapter 2), but there is no proof that this is the reason they have larger cheek pouches or are larger than females. Other researchers think that the size difference between males and females may be a matter of sexual choice, females preferring to mate with larger males.[23]

There is an ongoing discussion about how social or territorial orangutans are. This dialogue is important because if the social behavior of Sumatran and Bornean orangutans is different, it has implications for the debate about the genetic differences between the two populations (see Chapter 2). Some early reports[24] that orangutans are seminomadic, moving along stable routes and visiting different areas in different seasons, have been confirmed in recent years.[25] In the Gunung Leuser National Park, it has been noticed that some individuals stay in one territory more or less permanently, while others appear to move about.[26] There seems to be no fixed or standard way of being solitary or social if you are an orangutan.

Adaptability

One feature that describes all orangutans is their ability to adapt to different conditions and circumstances. Sometimes they are solitary, other times they are gregarious. Sometimes they stay in one territory, other times they migrate. Sometimes they are purely vegetarian, other times they eat meat. They seem to be expert in adapting to the vagaries of living in the rain forest. Such adaptability indicates high intelligence. As we describe in the chapters that follow, this intelligence is seen in their ability to learn and to solve problems.

Although orangutans are highly adaptable, this does not mean that they cope successfully with whatever hardships come their way. In Chapter 9 we show that they do not adapt well to living in zoos, and in Chapter 10 we discuss their inability to adapt to the destruction of their habitat. It is one thing to recognize their intelligence and adaptability and another to exploit that to the advantage of humans so that they can be looked at in captivity or in rehabilitation centers.

The Orangutan Brain

Not a great deal is known about the orangutan brain. Its overall shape and structure is very similar to that of other apes, including humans. All ape brains are smaller than human brains, but some of this is accounted for by their smaller body size. The volume of the human brain increased at some time in its evolution, although exactly when and how fast the increase was is now debated.[27]

An important part of the brain called the *neocortex* (also called the isocortex) evolved only in mammals, and it increased in size as mammals evolved into primates and then into apes. It is the largest part of the brain and can be seen from the top and the side. As the neocortex became larger, its surface became more folded (or convoluted). The ape neocortex is large and convoluted, although not as large as that of humans.

The front part of the neocortex on each side of the brain, known as the *frontal lobes* (Figure 3.1), is of particular interest because it has been associated with the planning of future behavior, decisionmaking, and artistic expression. Dean Falk of the State University of New York compared the size of the frontal lobes in chimpanzees and humans and found that relative to the rest of the cortex, the frontal lobes are larger in humans.[28] A team of researchers—Katerina Semendeferi, Randall Frank, and Gary Van Hoesen—at the University of Iowa used the new technique of three-dimensional reconstruction of images of brain scans

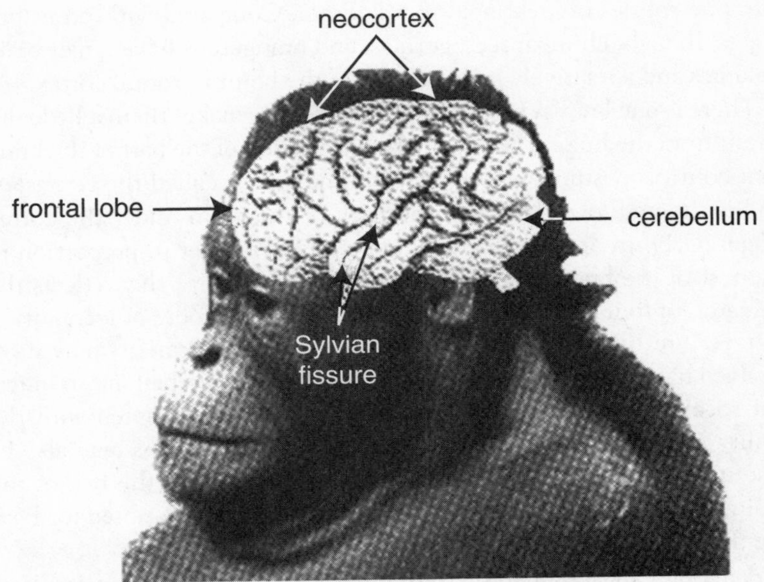

FIGURE 3.1 The orangutan brain, with some of the main structures labeled. SOURCE: *Based on Zilles and Rehkämper 1988.*

(obtained by magnetic resonance imaging) to measure the relative sizes of the frontal lobes in humans and apes. They measured the size of the frontal lobe relative to the rest of the cortex in one chimpanzee, one gorilla, and one orangutan and found no difference in size among these species, and no difference from the same measure in humans.[29] The varying results obtained by different researchers may be related to the fact that only a few brains were used to make the measurements and possibly also to the measurement technique used.

More research on more subjects will determine what is correct, but at present, we cannot say that the frontal lobes are the most important brain difference between orangutans and the other apes or between apes and humans. It seems that as evolution progressed, the size of the frontal lobes increased at the same rate as the rest of the brain, so the relative size remained unchanged. We cannot say that anything special happened in the evolution of the frontal lobes to make humans superior to apes. Nor is there any aspect of the frontal lobes that would separate one species of great ape from another. Compared with other primates, though, chimpanzees, gorillas, and orangutans have larger brain volumes and a relatively large portion of the brain is frontal cortex.

There is one known feature of ape brains that makes them a little different from the human brain, and that is the size of the part of the brain that controls posture and movement. The area is called the *cerebellum* and is situated lower down at the back of the brain under the cortex (Figure 3.1). In apes, the cerebellum is slightly larger in proportion to the rest of the brain than it is in humans.[30] This probably reflects the demands of living in trees. Brachiation and other modes of locomotion in trees are highly elaborate, and these accomplishments may have evolved hand in hand with an enlargement of the cerebellum. In different species of bats, a larger cerebellum has been associated with the ability to use many different ways of moving about. This may also be true in the apes. The cerebellum is largest, relative to the rest of the brain, in gibbons, and orangutans and these species are noted for having the most versatile ways of moving around.

A feature of the human brain that has received much attention is its asymmetry in the region of the neocortex that is used for speech and language. Most humans use regions of the left hemisphere to produce speech and process language. These regions are larger in the left hemisphere than in the right.[31] As a result of this size difference, one of the landmarks on the surface of the brain is found to be asymmetrical when left and right sides are compared. We refer to the Sylvian fissures, folds that run down and along the sides of the cortex (Figure 3.1). The Sylvian fissure is longer and lower on the left side than on the right side. The same asymmetry occurs in orangutans, as well as in chimpanzees and gorillas.[32] This is important because it suggests that all the great apes may possess those regions of the brain that humans use for speech and language. It is possible that they use this enlarged region in the left hemisphere for their own species communication. They may also use the same region when they are taught to communicate with humans using sign language (see Chapter 5).

Asymmetry of the brain in humans is reflected in our handedness. Most humans use the right hand to carry out fine manipulative tasks such as writing. Our own research has shown that orangutans are also handed.[33] They display left-handedness when touching the face to clean their eyes, ears, and teeth. Humans have a tendency to use the left hand for the same things.

Although little is known about the orangutan's brain, we can say that it is similar to the human brain in quite a number of important ways, but different in others. Orangutans have specialized adaptations for their lifestyle, as is the case for all species. They also have in common with us and the other apes brain features that indicate higher levels of thinking and learning (e.g., the large convoluted neocortex). As later chapters show, this capacity is manifested in the complexity of their behavior.

Part II

BEHAVIOR

4
MOTHERS AND INFANTS

Among orangutans, childhood is a distinct phase of dependence, a time when the offspring is unable to fend for itself and needs to learn the skills necessary for survival. Orangutan babies are vulnerable, just as human babies are. They need adult guidance and protection. Orangutans usually reach maturity when they are about fifteen years old—some may reach adulthood a little faster and others more slowly. They may "leave home" when they are about eight or nine years old, but for those first eight years of life, and especially the first five, they are dependent. During this long period of dependence, second only to humans, they form a strong bond, not with a group of relatives as do the other great apes, but with just the mother.

We have had the opportunity of seeing mother-infant interactions in free-ranging rehabilitating orangutans and always cherished these observations, particularly when we knew the individual well and had met her before pregnancy. In orangutans, as in humans and other species, motherhood brings with it definite changes in lifestyle.

We knew Jessica from several visits to Sabah Rehabilitation Centre, where we saw her grow up. By the time she was nine years old, she was depressed, listless, and inactive. She munched on leaves without conviction, spent hours investigating the food in her mouth, and then, quite often, spat out what she had chewed. Rarely would she play or swing on the branches. It seemed impossible to rouse her interest in anything. This was not how a nine-year-old orangutan should behave—she was just moving out of her teenage years and into the prime of life. But her life close to people, even well-intentioned people, had exacted a toll. Physically, she was healthy. Her ailment was psychological. She had no focus in life, drifted from one day to the next, and seemed barely conscious of other orangutans or of the people walking around beneath her favorite resting tree. She sat there sullenly for most of the day, staring into space.

It seemed like a miracle when we returned to the station one day and found a dramatic change in her posture and behavior. She had given birth, and the young one was a few days old. We watched her for hours from a distance. She was entranced by her youngster, hugging it, stroking it, examining the parts of its body, and supporting the back of its head as it suckled (see Figure 4.2).

Orangutan females do not commit themselves to eternal pair bonds or bonds in groups, as do chimpanzees, gorillas, and humans. Their relationships with males and the potential fathers of their offspring are quite fleeting, even if at times intense. When a female is ready and willing to mate, she consorts with an eligible male for about a fortnight and then their ways part. When she falls pregnant, the adult female becomes a single mother, without exception. No one knew exactly who Jessica's mate had been, but there was only one known adult male, Simbo, who occasionally still visited the rehabilitation area. He had once needed help for a fracture, over twenty years earlier. Everyone thought that he was the most likely candidate, as males do not like each other's company and will usually avoid overlapping territories or meeting each other.

Jessica's brief "love affair" contrasted strongly with what would follow—her time commitment in the future would be enormous. First, she would go through nine months of pregnancy, just as human females do. The orangutan is the only primate except for humans to carry a baby for that long. The only reprieve for her during that period is that subadult males, out for a first sexual fling, voluntarily give up pursuing a pregnant female. Second, there would be a very long commitment to her offspring once it was born (see below).

Nearing the end of her pregnancy, Jessica disappeared. Her usual resting and foraging spots were empty, and no one had seen her. In the rain-forest terrain, she could be anywhere, although normally she never strayed very far from the feeding platforms that were provided for rehabilitating orangutans. No one saw the birth of her offspring.

Toward the Birth

Very few people have known the moment of birth of an orangutan in the wild because the female retires to the treetops and gives birth high up in the canopy (often more than thirty meters up) in a nest she has built. During pregnancy she suffers aches and pains, especially in the last month, and this makes her stop and rest often. The closer it gets to the day of delivery, the more labored her movements become. And an orangutan's movements are rather different from those of a human. Imagine a nine-month-pregnant orangutan hanging from a branch high above the ground, swinging to reach the next branch without falling, calculating just how much weight that next branch will take at a time when she is a good deal heavier than usual.

In the last week before delivery, she stretches and bends, as if to overcome pain,[1] and begins to build resting stations in the canopy even during the day. She may spend long periods in these nests, which are surprisingly comfortable. They are built by breaking branches and arranging them skillfully across a fork in the tree. It takes a considerable amount of practice to be able to weave the branches in the correct way. The nest must not be flimsy or have breaks in the matting or the orangutan could fall through. It must support her fully (she weighs more than forty kilograms during pregnancy) and, after birth, must be firm enough to prevent the newborn from slipping through a crack. The nest is big enough to allow the adult orangutan to stretch out fully. When fully erect, Jessica is just over one meter in height, so her nest would have needed to be nearly one meter in diameter and constructed so that she could put her legs up and rest her back.

From the forest floor, one can see only the underside of the nest, a large tangle of leaves and branches. The layers of the nest are so thick that it is impossible to distinguish the contours of the orangutan's body. The only sign of occupancy is an occasional wave of a hand, casually drooped over the edge.

When the day of the birth arrives, the female usually does not leave the nest. She will not eat, and she will be entirely on her own. Almost all births happen under cover of darkness, an evolutionary precaution, we think, that helps the female to complete the birthing process with fewer threats to her life or the baby's. What these threats might be is not entirely clear. There used to be dangerous predators in orangutan territory, such as the clouded leopard, but such predators are now very rare. Clouded leopards hunt at night and are agile in trees but usually do not go as far as the upper layers of the canopy. Snakes can be a risk to newborns of many species, and a number of them are nocturnal and hunt in trees. Avoiding snakes, therefore, is unlikely to be the reason for giving birth at night. Perhaps the real risk of a daytime birth could be the vulnerability to birds of prey. They may find a newborn orangutan a welcome snack.

Jessica probably built her nest as high as possible, out of sight of everyone, especially humans, and gave birth at night, silently but not without pain. Orangutan labor and the length of time it takes to give birth are very much like the human experience. Births have been observed in zoos,[2] and there is one wildlife record.[3]

One day Jessica returned to the feeding station with a little infant in her arms. It looked healthy and so did Jessica. The most powerful image that we retain from that day is her smile. The glum, empty stare and the emotionless expression had vanished from her face and had been replaced by an almost constant smile and a gentle manner of attending to her youngster (Plate V; Figure 4.2). Even if we cautiously refrain from assuming that orangutan behavior resembles human behavior, it was clear that Jessica greatly admired her offspring. She was very involved in her new role and fixed all her attention on the newborn. Her smiling suggested that she was a proud and contented mother.

To our eyes, very young orangutan babies are comical rather than appealing (Figure 4.1). They are usually scrawny. The eyes are surrounded by a light circle of skin, giving them a clownish appearance, an impression made stronger by the sparse hair sticking straight upward on the head. The face may be covered in wrinkles, and the toothless mouth moves in uncoordinated ways.

It was just as well that Jessica was so enamored of her new youngster (such intense involvement being typical of orangutans in the wild) because she was in for the long haul. There is no state of motherhood among primates, even humans, that is so all-embracing, so utterly complete and singular as among orangutans. Once the young is born, orangutan mothers do not seek contact with males and may ward them

FIGURE 4.1 *Jessica's baby, six days old. The first active attempts at climbing get the infant as far as the mother's shoulders.*

off, even mature adult males. They direct all their attention to the care of the infant for the first five years after its birth, and for part of that time, they carry the offspring on their body.

Good and Bad Mothering

How do orangutans know how to care for their babies? Some decades ago, it was still believed that maternal behavior was instinctual, induced and mediated by hormones. The idea that good mothering is the result of experience and learning took some time to be recognized, even for human mothering.[4] This shift in thinking raised many new questions, particularly in regard to great apes in captivity. If knowing how to care for their offspring is not innate, how do they learn?

In the natural environment, there are three main ways in which the female orangutan can learn how to raise her offspring. If she is lucky, she may see her mother raise another infant. By then she would already be five to eight years old and would be able to observe her mother interacting with her sister or brother. But opportunities for an orangutan to learn about maternal care by watching her own mother are rather limited. We know that the birth intervals of the other great apes are relatively long for mountain gorillas, four to five years, and for chimpanzees, six years.[5] The birth interval of orangutans is even longer, at about eight years,[6] which means that orangutans have an extremely low reproductive rate. Females in their natural habitat raise, at best, four offspring in their lifetime and often just two. A mother will usually not consent to sexual intercourse with a consorting male and the chance of becoming pregnant until her offspring is at least five years old. As a result, a young female might get one chance to watch child rearing by her parent, but many will not have this opportunity.

A second method of learning to rear offspring depends on a few brief encounters with other adult females and their offspring. The brief social gatherings that occur when favorite fruits are ready for picking means that several mother-child pairs may come together for the time it takes to strip the tree. While the mothers are feeding, their offspring have the chance to observe them interacting with the younger orangutans.

Females who lack the opportunity of observing mothers rearing young must raise their own offspring by trial and error. The mother learns as the offspring learns. Primate literature has long since established that in many species, primiparous mothers (i.e., first-time mothers) show few mothering skills and are more anxious than multiparous mothers (i.e., those who have reared several young).[7] However, if all orangutan mothers with firstborns were poorer mothers than those that had given birth to several offspring, this would suggest that the majority of orangutans ought to be poor mothers, and there is no evidence to support this in wild orangutans.

In healthy wild-born populations, there is no evidence that primiparous orangutans make poor mothers (despite research evidence of this being the case in other ape species), but there is substantial evidence that a female infant removed from her mother before the required learning has been completed will have severe psychological damage and be unable to mother effectively as an adult. Researchers of captive primates have found that once mother and infant are separated, interaction between infants and juveniles increases initially but then decreases sharply, while depressed behavior is initially low and then increases significantly.[8] Some researchers argue that the mother offers a secure base for exploration, and if this base is taken away, the infant or juvenile becomes anxious and avoids contact with anything new and unfamiliar.[9] In addition, separation from the mother interrupts the gradual learning of skills at whatever point the separation occurs. This may impair learning of appropriate foraging behavior and adequate locomotion in trees, as well as mothering.

Poor mothers exist in orangutan society, although they are not the primiparous ones. Some, especially those who have grown up in captivity in a zoo, may have suffered traumatic events in their infancy and therefore tend to become unstable and less capable mothers. Some do not know what to do with the newborn. There are examples of incredible cruelty in captive orangutan mothers pulling and twisting the limbs of the young one or biting hard into the small fragile arms or even into the face, pushing the baby away when it seeks the nipple to suckle, or dealing it painful punches. We know of one case where the mother wore her youngster like a hat. She also decided to use it as a cushion and crushed the baby under her weight. Many infants die as a result of poor mothering in captivity.

Anne Russon of York University, Toronto, tells the story of a rehabilitating orangutan who was cruel to an infant, obviously unaware of what she was doing. An adult female called Supinah at Camp Leakey (in southern Borneo) had managed to get hold of a battery and seemed intent on breaking it open. She needed a tool to help her crack the casing so she grabbed one of the orphan infants, held him down, and used his head as an anvil for the battery.[10] Studies of infant abuse among primates indicate that early social deprivation and poor treatment lead to abusive or neglectful parenting, just as in humans.[11]

Despite adequate mothering by first-time orangutan mothers, there is room for improvement, and experience tends to help with the rearing of subsequent offspring.[12] But reports of good and bad mothers are not confined to orangutans. There are similar reports of inadequate mothering among other great apes in captivity, and a good deal of money and time has been spent in the recent decades to try to improve the situation. In response to research findings, zoos are running programs to teach primates appropriate maternal care.[13] Fortunately, orangutans are fast learners. Usually, all that is needed is to let the female watch the mothering of an infant by another orangutan female, and the inadequate mother can turn, almost overnight, into a very adequate mother, unless she has other behavioral problems.

Motherhood Duties

Motherhood duties for an orangutan can only be described as arduous. The female has no relatives around to help her or take the infant from her, even for a short period of time. She has no one to help her gather food or find shelter. The mother is the baby's only means of transport and support, its only source of food, comfort, and safety, and often its only source of information and essential learning experiences. She is the sole caregiver, and the infant is dependent on her abilities and attentions to an extent rarely found in other species. In no other species of primate, except humans, is parental supervision and care spread over such a long period (solidly for five years and then another two or three years of adolescence) and left so entirely up to one adult. "Mother orangutan" is the primeval symbol of motherhood.

We usually distinguish three phases of maternal behavior. The first is a time of support and solicitousness for the infant. In the second phase, the mother becomes ambivalent toward the infant's autonomy and takes a little less notice when the infant begins to stray from her body. The third phase is a period of weaning, of the mother's active rejection of the offspring. These stages in maternal care last three to six months for lower primates (e.g., prosimians), six to twelve months for monkeys (many species), and four to six years for apes,[14] the longest of which is for orangutans.

The orangutan mother ensures that the infant is attached to her body at all times in early infancy. The infant needs to hang on tightly because the orangutan mother may be perched precariously thirty meters above the ground. One loosening of the infant's grip on its mother's hair and the infant would tumble to the ground and be perhaps maimed or killed. However delicate the infant's hands may look, their grip is extremely firm and the infant can support its own weight purely by using its hands. The legs begin to play a role when the infant is a little older. They are folded around the mother's body as far as they will reach, and the feet, which are as agile as the hands, can then also grip the mother's fur.

Slightly older infants seem to determine their own place for gripping the long hair of the mother's body. The mother shows no particular preference for where on her body the infant should grip, at least not when she is stationary. When the mother is about to move on, she may scoop up her infant and place it in a position that is comfortable to her. We have observed this scooping motion on several occasions, by three different females with infants under eighteen months of age. After being positioned, it is up to the infant to hold on.

Some orangutan females do not have very much body hair, and we have seen cases where the infant had to maintain its hold by pinching the mother's skin tightly. Mothers are usually very tolerant in accepting the pain inflicted by the infant. At best, they might take one little hand and shift it to another position on their body. The infant is usually cradled on the mother's chest when very young and then graduates to clinging to the side of the trunk or the hip. When the mother descends a tree backward, the infant rides on the lower part of her back, and when she is climbing upward on a trunk, the infant may move higher up on her back.[15] When the mother begins to move through the trees, we found that the most common position for the infant is on the hip, sideways, so that the mother is free to negotiate branches and move her legs and arms.

Mother-infant interaction, while not demonstrative in most cases, is nevertheless intense in orangutans. The relationship between infant and mother is generally very affectionate. As in humans, there is a novelty effect with a new birth, and the female is at first very curious about every part of that little body in her arms. Some new orangutan mothers spend hours examining the body of their infant. Jessica was particularly fascinated by the hands of her new offspring. She often took each of the tiny fingers between her thumb and index finger and looked at them in turn. Sometimes she rubbed one of the baby's fingers gently. Most of this behavior tends to disappear over time, hardly surprising given the long development time of the infant.

Harrisson observed that the infant communicates affection to the mother in a number of distinct ways. It may expose its gums or teeth (i.e., smile), open the mouth wide, "bite" the mother's face gently, and lick the mother's lips and mouth occasionally. Another affectionate behavior is "combing" of the hair and limbs with the mouth and fingertips.[16] The orangutan mother will sometimes respond to the infant's movements and signs of affection by hugging and cradling (especially very young infants).

Given that orangutan infants have little opportunity to make friends, do they engage in play? In play, infants are said to develop the skills that they will need later for survival.[17] Wild orangutans do not play together often; compared with chimpanzees and gorillas, the occurrence of play among orangutans is rudimentary.[18] The reports that we have about play between orangutan mothers and their infants are largely contradictory. Some writers have found that the females play only rarely.[19] MacKinnon, however, has described some play behavior of mothers with their infants, including tickling, biting, poking, and gently knocking the infant with closed fists.[20]

We have watched mothers and seen that some play more than others. Some mothers tickle their young and engage in raucous jostling matches, others (less often) take to grooming. Many infants use their mother's body as a playground. It depends a little on the mother to indicate what parts of her body may be used in this way and what parts can be inspected or are off-limits. We once watched a male infant of about eighteen months freely examining his mother's genitals for a leisurely period of half an hour. The mother made no objection. There may well be individual differences. Orangutans are, after all, sophisticated primates and not just their prior experience but also their personalities will determine what kind of interaction occurs. Some mothers may allow more freedom of movement to their youngsters than others.

Jessica engaged in very affectionate behavior (Figure 4.2). She kissed and hugged her newborn, pressing it gently into her arms. Her behavior toward her infant looked very promising, and in all likelihood, Jessica would be a good mother.

Growing Up

Recognition of the human infant as a complex and sophisticated organism is a discovery of the twentieth century. William James, one of the founding fathers of psychology, still thought that infants were "insensate" and their sensory perception scrambled and confused.[21] We now know that this is not the case. Infants are capable of processing very complex information,[22] and further studies of primate infants have confirmed this view. Indeed, studies on newborn chimpanzees have revealed that they direct sustained attention to visual and auditory stimuli.[23] It has been found that social stimuli (e.g., the face of the mother or her sounds and gestures) are attended to more than nonso-

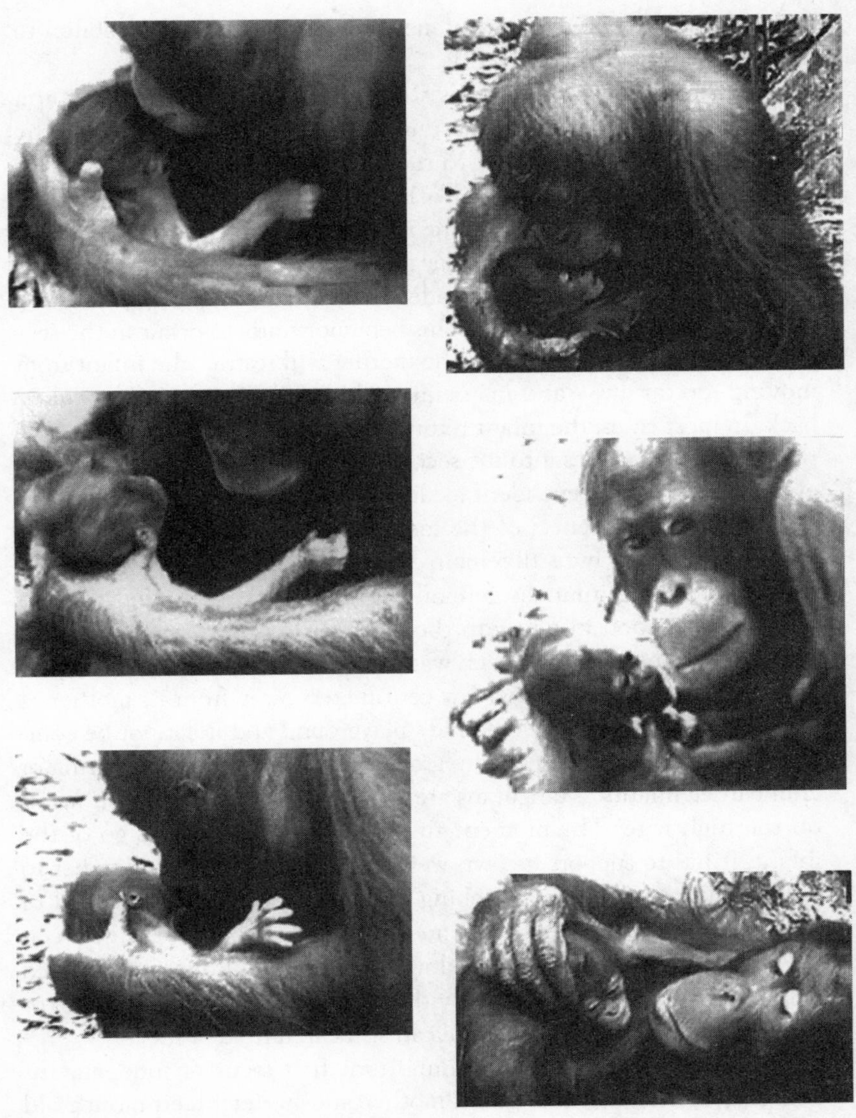

FIGURE 4.2 *Attentiveness and great tenderness is shown by this mother toward her infant. Jessica's demonstrated abilities at mothering may suggest that she has been able to observe mothering in her past.*

cial stimuli. Humans are not unique among primates in their ability to engage in face-to-face contact.[24]

Jessica's six-day-old infant was still rather uncoordinated in its arm movements. The only two activities apparent from birth are the infant's need to find the teat to suckle and the ability to grip firmly with the delicate little (humanlike) hands. In the first year of life, the orangutan mother ensures that the infant is constantly attached to her body.[25] She also makes certain that the infant remains close to her during the brief periods of independent investigation that are allowed swinging, playing, or climbing. This behavior starts to occur in the second year of life, but even then, the mother will restrict the infant from moving too far away and may emit a deep belching sound to call it back. In most cases, the infant returns instantly.[26]

As the infant moves into the second year of its development, there is evidence of some independent feeding.[27] With these foraging or playful excursions, the closeness of the mother's interaction with her infant diminishes slightly but still remains intense.

Feeding by the infant is generally strictly supervised, and the offspring will respond instantly to the mother's minutest signal and rush back to her. The reason for such watchfulness is simple: Even when the orangutan is "playing" only a few centimeters away from its mother, it does not have the safety of a child's playground and it cannot be compared with the frolicking on the ground that is customary in gorilla or chimpanzee infants. Orangutans are trapeze artists and take their meals on the high wire. The moment the orangutan mother lets go of the infant, it has to support its own weight on a branch and remember to hold on. Its swinging and climbing location may be the equivalent of the top of a four-story building, and there are no safety nets between the infant and the ground far below. The first small distances created between mother and child are within reach of the mother's arm (or leg). If the youngster tumbles, she can still catch hold of it.

We were able to watch the important first steps of independent swinging between an orangutan mother and her eighteen-month-old offspring high up in the canopy. The youngster was dangling precariously from a branch, but very close to the mother's face. Clementine, the adult female, began feeding without sharing any with her offspring, who was watching intently. Eventually, the infant took one hand off the branch and reached for the food in her mother's mouth. Clementine turned her face away and the youngster followed her, finally managing to snatch a tiny piece of fruit from her mouth. By now, the youngster was dangling by one hand, using the other to hold the food. The infant then began to consume the food and thus ate at the same time as Clementine. These are instances of first "independent" feeding. The offspring has to ask for the food, reach for it, hold it, and take it to its mouth.

Interestingly, this infant followed its mother's example in a specific way. While the mother fed with the right hand, the infant fed with the left, facing the mother and watching her. Here is a mirror-image situation (Plate VI), and we suspect that learning in mirror-image fashion is perhaps the first attempt at imitation (see Chapter 5).

By the age of two years, the young orangutan moves and feeds independently of its mother, although some suckling may continue well beyond the second year. Captive orangutan offspring will often suckle up to five years of age. There may be times when the infant needs help in climbing or getting to another tree, and in those cases, the mother gives assistance.[28] We have observed situations where the mother uses the infant to reach for a branch that is out of her reach. The infant becomes an extension of her arm, and surprisingly, the infant knows that it is meant to hold onto the branch and pull it toward the mother.

Even after the age of three years, the juvenile, now quite capable of independent locomotion, follows the mother by holding on to the hair on her rump.[29] In this manner, orangutan infants slowly learn to judge distances.

In general, juveniles of the age of three years or older appear to be very confident, but this is only the case when the mother is present. They need reassurance that she is there. Such reassurance may no longer require touching but involves some other form of communication.[30] This need for contact reveals not only the mutual reliance on appropriate signals by mother and offspring but also an underlying anxiety by the offspring. The juvenile is anxious about the possibility of being separated from its mother. In the first years of life, it is the mother who shows anxiety and concern about the infant moving away from her, while in juvenile life (two and one-half to five years), it is usually the youngster who seems anxious.

The smallest movement away by the mother may cause the young one to scream and run or climb after her. This behavior intensifies as the mother slowly begins to wean the youngster, not just from suckling but from her constant attention. We have seen adolescents of five to eight years of age throwing temper tantrums and sulking when the mother failed to pay attention. The pinnacle of injustice done to the youngster seems to be the mother's decision to throw the juvenile out of her sleeping nest. Very dramatic scenes can ensue where the scenario is almost always the same. The mother says: "You build your own nest and stay out of mine," and the juvenile responds by screaming, thrashing branches, crying, and trying to get into the nest. For several weeks, the mother will get very little rest because her offspring does not give up.

Fractured Childhoods

We humans think that childhood is meant to be a happy and carefree time, for human and animal infants alike, and we usually have no problem transferring our expectations to the great apes. Childhood is normally a time of carefree play, with a reliable food source and constant protection and learning.

Not all orangutan children are so lucky. Orangutan babies, while sometimes scrawny, can also be extremely cute, particularly when they are a little older. They are responsive and entertaining and readily remind us of human babies. Orangutan babies have always held a special attraction for humans that seems to be universal, and it is this quality that has persistently been the cause of the greatest tragedies in orangutan society. During the nineteenth century, hundreds of orangutans were shot and shipped back to museums in Europe, but the twentieth century has seen a different market emerge. While elephants have been killed for their ivory, rhinos for their horns, tigers for their pelts, and gorillas for their feet and hands, orangutans have been hunted and traded internationally for their live babies. This happens less often now than in the past, but there is still an illegal trade. In order to get a baby, the mother has to be shot—only then can the little one be collected safely. While the infant orangutan is small, it can be kept like any other pet. Some are held in cages, while others have to work, either as an attraction in a zoo or in a nightclub or circus.

So sought after were orangutan infants that at the local level in Indonesia and Malaysia, they were considered valuable presents by officials of army, police, and even government. Orangutan babies were presented as awards and signs of appreciation to others.[31] Much of the international trade disappeared in the late 1990s because of stronger international interest in the fate of orangutans, but in parts of Malaysia and Indonesia, baby orangutans are still being confiscated regularly from private homes and remote villages. Most are orphans because their mothers have been shot in plantations or found electrocuted or killed accidentally during tree felling.

It is difficult to imagine what the loss of a mother means to an orangutan youngster until you have met one and seen the anguish associated with such a loss. We met one orangutan personally—Abbie, a young female of no more than three years of age when we first met her. Apparently her mother had been electrocuted. At that age, orangutans are not able to fend for themselves and do not know enough about their world to survive on their own.

Abbie was a lost soul. She wandered around with other orangutans at the Sepilok Rehabilitation Centre in the days before mass tourism had discovered this outpost. We stumbled upon her quite accidentally and crouched down to speak to her. At first, she made a few hesitant advances, then she clambered up from behind and occupied Gisela's body in piggyback position. Abbie folded her legs around Gisela's waist, closing the circle with her entwined toes in front, and folded her arms gently around her neck. She changed position a few times, eventually resting her head on Gisela's shoulder (Plate VII).

Abbie was heavy, but all the same, Gisela carried her around while Abbie made little purring sounds like a cat. Her hands felt warm and pleasant. Orangutan skin, at least that of young orangutans and usually all females, is very soft and delicate, rather like that of an aged person. Abbie's body smelled nicely of an orange-flowery scent. Her facial expression changed during the afternoon. She seemed sad at first, but by the end of the day, her head was held high and she clearly had no intention of leaving, so strong was her need for closeness and protection.

Of course, the time for separation came. We tried to heave her off Gisela, but she clung on hard. When, after much struggling, she was finally removed from her seat on the hip, she held on to Gisela's legs with a firm grip and an increasingly alarmed expression. Then she started screaming. The screams were spine-chilling and could only be interpreted as cries of separation and desperation. A Sepilok worker came and wrestled Abbie away from us. She threw a temper tantrum and screamed continuously as we walked away quickly. With hindsight, our actions had been as unforgivable and irresponsible as they were innocent. We had playfully enjoyed the encounter, but there had been nothing playful in it for Abbie. She had desperately sought, and thought she had found, a new mother, only to be separated a second time. We could have infected her with our human diseases. We certainly caused her misery. But the encounter ensured that from that moment on we would speak and write on behalf of orangutans. One juvenile had changed our lives forever. This crucial encounter began our research on orangutans.

Thankfully, Abbie's story has a good ending. She has been a rehabilitation success, as far as she could be in her circumscribed environment. We continued to return to Sabah in the years following this encounter, thinking especially of Abbie. We always found her again. Who can tell, but every time she was bigger, looked healthy, and seemed more confident. Two years after our first encounter, we finally spotted her after three weeks of daily forays into the forest. The tourists had just left, and we were alone. We called out to her and then watched as she began slowly and deliberately to amble toward us. Gisela had sat down on the forest floor, and when Abbie reached her, she sat down opposite her. A few greeting silences passed, usually with eyes averted, but Abbie remained seated close by. Inexplicably, she eventually took Gisela's hand, turned its palm upward, and ran her index finger over the lines on the hand. Then Abbie and Gisela sat still for a good while, literally just holding hands.

Eventually Abbie got up and disappeared, only to return a little later with another female juvenile her age. As far as one can speak of friendships among orangutans, Alice was certainly a friend to Abbie, or at least a good playmate. The two of them gamboled about in front of us, barely fifty centimeters away, and showed us all manner of tricks. Then Abbie got up and, gravely, took Gisela's hands again, just holding them (Plate VIII). When she let go finally, she slowly turned around and walked away.

On another visit, several years later, she meandered toward us again and sat down near us but no longer sought to touch us. This time she had another companion. Almost old enough to be a mother herself now, she had instead adopted a little three-year-old male (Plate IX), or had been adopted by him. Tom had lost his mother in circumstances similar to Abbie's, and he could not be separated from her. If she turned away, even very slightly, he clung to her anxiously, and when she wanted a few minutes to herself, he shrieked in absolute terror. Tom was not the kind of charge to allow Abbie a quiet, contemplative meeting with us on the jungle floor, but she stayed close, about half a meter away, looking intently at Gisela.

Not all stories end that well. Other orphans die, like the skinny young male Gemasil, who became more and more depressed (see Figure 7.4). He had found a companion briefly (a subadult male), but eventually that companion deserted him. Others suffer permanent psychological damage and skill deficiencies. Zoos are still full of orangutans with severe stereotypical and abnormal behaviors. Worse, of the many that have been taken from the wild, only a fraction survive the ordeals of illegal transportation or the shock of translocation. Any two orphans together offer a heartrending sight. They huddle in each other's arms, groping for the security they have lost, their eyes portraying an image of innocence mingled with helplessness and fear.

Our initial experiences of physical contact with just one orangutan infant made it difficult for us ever again to ignore the closeness between the human species, *Homo sapiens*, and *Pongo pygmaeus*, the orangutan.

5
LEARNING

It is easy to say that the mother-infant relationship is the basis for learning and skill development in orangutans. It is much more difficult to show how young orangutans actually learn and what the processes are that will eventually turn a youngster into an effective and competent adult.

Orangutans, together with the other great apes, are probably the most sophisticated of the primates, and they seem to be cognitively more advanced than the monkeys, although this is not absolutely certain.[1] We now know something of what they can do in the wild. There is also a good deal of information available on specific orangutan behaviors that seem to occur only in captivity or when there is some other form of human contact, but they are impressive all the same. If an orangutan can build bridges, paddle boats, and make fires, there must be something special in its makeup that enables it to do so. It is a capacity to learn, watch, adapt, imitate, and acquire skills that other species have not been able to acquire.

Learning by Observing

We do not know precisely how learning occurs in the orangutan infant, but it seems reasonable to assume that there is relatively little difference between human and ape learning in infancy. The infant, human or nonhuman, is not on its own. It has at least one adult to refer to, an adult that is also the food source and provider of all the other facets of comfortable life as well. As long as the infant's bodily needs are met and it is protected from harm, it is free to watch the world and perhaps learn by observation. Putting the learning into practice is dependent on physical maturation and on social conventions (or rules) within a species.

Primate infants in general are regarded as skilled information gatherers.[2] They learn by observation. When orangutan infants are a little over a year old, they begin to take a keen interest in the food the mother is consuming. We have seen youngsters aged thirteen to eighteen months begging food from their mother, from either her hand or her mouth. They may try to take the food into their hands but, in the early stages of development, will seek to lick or bite off the food directly from the mother's mouth. Usually, the mother restricts access to the food she has collected. We have not once seen a mother actively offer food to her offspring. The offspring either suckles or obtains food by taking some from the mother or directly from the source.

Despite these restrictions (which may exist for a good reason), the infant learns to recognize acceptable food by sampling most of the foods the mother eats. As the infant develops a taste for other foods (other than suckling milk), it begins to follow the mother's hand to see where the food is coming from and so gains knowledge of the food source. Exposure to a rich variety of foods over the seasons and years slowly acquaints the youngster with the very important fact that not every type of food is available at all times and that sometimes it is necessary to travel to a specific destination to obtain the food. This is particularly true of fruit (see Chapter 3).

Attentiveness and Eye Gazing

Learning by observing has a number of interesting twists in the orangutan. In Western human cultures, we tend to associate learning with attentiveness and attentiveness with eye contact. Direction of eye gaze is also important for learning—if the skill to be learned is not watched closely, how can it be learned? Gary Shapiro of the Orangutan Foundation International attempted to teach four orangutans some sign language.[3] In the learning phase, he noted that the orangutans averted the direction of their eye gaze and appeared not to be attending.[4] The fact that they did learn to use sign language showed that they really were attending, although more furtively than other apes. Orangutans do give the impression that they are not looking, watching, or being attentive. Human observers may form the view that the orangutans are not interested or curious and are therefore unlikely to succeed in a task that depends on observation by the learner.

But is this judgment of orangutans correct? First, as this chapter shows, orangutans are impressive in the speed with which they learn and solve problems (see also Chapter 6). Second, we must always allow for the limitations of human perception. Until quite recently, humans assumed that all matters of animal perception were detectable by human observation. In other words, we assumed (incorrectly) that our perceptual framework was sufficient to detect, assess, and describe all that went on in the perceptual world of animals without realizing that some animals' perceptual capabilities may be superior to our own or, at least, different from our own.[5] This attitude toward animal perception has changed dramatically with the use of improved technology. We now know that we may not see, hear, smell, or sense as much as some animals do (in regard to ultrasound, infrared perception, and perception of magnetic fields and electro-impulses).

Our own research[6] produced over seventy hours of video footage on orangutan behavior, and we were able to extract examples of eye gazing by analyzing the videotapes frame by frame. Our initial impression was identical to Shapiro's observation of orangutan behavior. It seemed that orangutans avert their eyes constantly and so, presumably, have little time to absorb new information. After frame-by-frame analysis, we identified more than 200 specific instances of looking, of which only 3 percent were mutual eye gazes (both animals looking at each other). Orangutans, it seems, do not make direct eye contact with each other very often. In Western human cultures, direct eye contact is highly valued. It may not be polite in orangutan society, as indeed it is not in some human cultures, but such conventions are not, in themselves, an indication of whether attention or learning is occurring.

There is also the impression that orangutans do not look intently or long enough at an object or another orangutan or human, even without mutual eye contact, to benefit from what they are seeing. Our frame-by-frame analysis showed that this impression is, in fact, incorrect. We found that the orangutans were looking at each other, not directly, but by glancing sideways.

The reason we cannot easily determine whether the orangutan is actually attending is because, for most of the time, their heads do not face in the same direction as their eyes. An orangutan may have its head turned to the left, seemingly indicating it is looking to the left, but instead, it may be glancing back to look sideways at another orangutan or a human. They often cast repeated furtive glances in a direction that is not easily betrayed by the position of the head or, for that matter, the whole body. Particularly from some distance away, it can be very difficult to judge what an orangutan is looking at. It is made harder by the fact that the glances toward an object may last each time for only a second or two, although over a span of time, they add up to a good deal of looking. From our own observations, this furtive looking seems to be a pronounced trait in both subspecies of orangutan. We have seen this behavior in free-ranging Bornean orangutans and in captive Sumatran orangutans (Plate X).

So although their eyes do not meet, orangutans do look at each other and use such looking for a variety of functions. On one occasion we became separated in the forest. Lesley and the workers of the Rehabilitation Centre were far away, and Gisela was left alone. All the orangutans had disappeared, and she was about to leave when she saw in the distance the unmistakable shape of an adult male orangutan, well-known Simbo, of the Sepilok Centre. She prepared her video camera and filmed him in the distance. Suddenly, his measured movement changed pace. He swung very rapidly from tree to tree directly toward her (orangutans can move with lightning speed through the forest). Within minutes he had reached her, stopping short two meters away and just three meters off the ground. He was close, uncomfortably close, and he was thought to be dangerous. There was no way Gisela could outrun him, and the opportunity to meet him face-to-face proved too much of an incentive to contemplate departure. Gisela looked down and silently praised the modern video camera for being noiseless and allowing her to watch Simbo with "eyes averted," through the tilted viewfinder (Figure 5.1).

Simbo had come over to inspect this human for reasons of his own, but during his entire "examination," which seemed to last for an eternity (about fifteen minutes), he did not seem to look once at Gisela.

FIGURE 5.1 *Simbo approaches (a) and pretends to look elsewhere (b). His eyes move continually (c), but the video camera manages to capture a split second when Simbo looks at Gisela (d).*

Watching him through the viewfinder, she did not once detect him staring or even glancing at her, although the nearness of his impressively large body was unnerving enough. She stood completely still, trying to ignore the pearls of sweat trickling into her eyes. Simbo hung almost motionless on the trunk of a tree, only his head turning from side to side as if leisurely surveying his terrain. When he finally decided to leave, his departure was unhurried, even hesitant. Gisela remained motionless until there was a safe distance between them and she was able to move.

Later analysis of this film sequence showed that Simbo had constantly inspected Gisela with sideways glances (as also seen in Plate XI), at times when he appeared to be looking far away and in a different direction. At the end of the encounter, he had even given her a number of "up-and-down" looks, which the poor human had missed entirely. These were presumably not friendly looks, and may have issued a warning. We know that in gorilla society, the glare of a male will usually suffice to change the behavior of his family. And there have been a few documented cases of male-male encounters of orangutans that employed direct eye gaze.

This behavior of direct gaze avoidance and apparently limited visual attentiveness can be interpreted in a number of ways, ranging from deception to cultural coyness and social rules. For instance, it would be an act of deception if I pretended to look at something over to my right when I actually had my mind on the fruit in your hand. I would snatch this from you the moment you also looked in the direction you thought I was looking.

Leaving social rules and other forms of communication aside (see Chapter 7), in the context of orangutan learning our findings suggest that our perception of attentiveness is superficially incorrect. It is derived from our own cultural dictates on attentiveness and politeness (look at me when I am talking to you) and also from our limited ability to perceive difference. We obviously find it difficult to detect diverted eye movements, and we are confused by the noncorresponding movement of head, body, and eyes. Only the video camera catches an accurate record, and by detailed analysis of the videotapes, we can reveal the full sequence of gazing patterns in orangutans.

It seems that orangutans—whether juveniles or adults—see and watch everything while pretending not to see, watch, or care.

Learning by Imitation

Observation alone is not enough, however, to equip the young orangutan with the skills it needs. Eventually, the orangutan infant will have to try its own hand, and these attempts at self-management may be learned by imitating what the mother does. This is not always easy. The two main areas where imitative learning may be involved are in collection of food and movement through the trees (locomotion), although there are many other situations, including nest building. Food and locomotion have very high survival value, so it is of vital importance to learn about them.

Imitation has been a topic of debate in ape research for some time, and it has been reinvigorated in the 1990s. This widespread interest in imitation, and in the imitation of tool use in particular, is connected with the evolution of *Homo sapiens*. We want to know how we, as a species, evolved into such an extensively tool-using group. We know already that children imitate.[7] Hence, we look to the great apes and monkeys as our nearest evolutionary kin to see how far these abilities can be traced and whether there is a difference between monkeys and apes.[8]

We do not know exactly what processes are involved when orangutans acquire skills such as food selection and climbing, but there is increasing evidence that in addition to motivation,[9] imitation learning may be involved. Several writers point out that this imitative capacity qualitatively separates the great apes from monkeys and that, therefore, imitative behavior is interesting from an evolutionary point of view.[10]

Most monkeys seem to have great difficulty imitating behavior even when they have been taught to use tools in a number of trials.[11] By contrast, there is increasing evidence that great apes imitate behavior. Byrne speaks of something like a "technical intelligence" in the great apes, which becomes evident in the imitation of manipulation and locomotion and also in tool use.[12] This suggestion is based partly on the fact that chimpanzees use tools in the wild[13] and orangutans also use tools extensively, particularly when in contact with humans.[14]

Tool Use as Imitative Learning

Free-ranging (i.e., ex-captive and rehabilitative) orangutans make use of a wide variety of tools. They use sticks or planks to dig holes in the ground and break open objects. They use sticks to hit other orangutans, to pry off loose objects, to break open fruit, to reach for the branch of an adjacent tree when locomoting. They also climb onto and balance on sticks when playing or use them as a ladder to reach a window, to stir liquids, to remove biting insects from the hair, and to scratch themselves.[15] These kinds of tool uses could be learned by watching humans nearby or they might be related to tool use seen in wild orangutans (see more in Chapter 6).

Here, we want to concentrate on tool use that has obviously been learned by observing humans and to highlight the aspect of learning as imitation. This occurs particularly in rehabilitation centers that have allowed the orangutans to roam free and mingle with people. For instance, in Camp Leakey in Tanjung Putting (Kalimantan Tenga, South Kalimantan, Indonesia) orangutans have learned countless tool-using skills.

Anne Russon of York University, Toronto, tells the story of a subadult male called Apollo Bob. He was an ex-captive, slowly being rehabilitated, but he was found to be quite a handful at Camp Leakey. All camp buildings had to be guarded because orangutans like breaking into houses and sometimes vandalized them. A member of staff on guard duty at one of the houses took up playing the camp's one guitar to relieve his boredom. Apollo Bob soon joined him on the verandah and then disappeared under the house. After a while the guard went to the kitchen for a cup of coffee, leaving his guitar propped on the porch. When he returned, the guitar was gone. He asked everyone in camp, but they all denied having taken the guitar. One of them suggested that Apollo Bob might have it. Unconvinced, the guard went back to the house to look for the guitar again but found nothing. Then he heard a plunking sound in the distance and followed the sound to the visitors' bunkhouse. Beneath the house was Apollo Bob, holding the guitar in one hand and plucking the strings with the other. We are not told how good the guitar playing was but, clearly, Apollo Bob had learned the first rudimentary lessons about the guitar—how to hold it and what to do with it—by imitation.

In all, Russon and Galdikas identified more than 300 cases of imitative behavior in the camp.[16] There were, for instance, orangutans in camp who had seen workmen bridging the river with logs before they adopted the behavior themselves.[17] Other examples of imitation include siphoning fuel from drums into cans, imitating the cook's use of fire by fanning embers with a lid and placing a cup of liquid on the fire, imitating the gardener weeding, and imitating painting of the floor and buildings.[18] The camp is full of such stories, and some of the most astonishing have been featured in documentaries, such as orangutans washing clothes with soap by the riverside and setting off in a boat, equipped with paddle.

The question is not just whether some simple or even more difficult imitative acts are completed correctly but whether the action has been understood. Supinah, an adult rehabilitating female, took a particular fancy to sharpening the blade of an ax, and she also learned that the stone needs to be wetted first before the sharpening is done. Whether she had understood conceptually why the stone needed wetting is not clear. Imitation alone might be sufficient in some activities to achieve the desired results. Even though it has been demonstrated that imitation plays a major role in tool use by apes, some of their tool use is acquired by individual experimentation.[19]

Learning to Climb

It is not surprising that one of the first skills an orangutan learns is how to climb and move, first on its mother's body and then through trees on its own. Knowledge of traveling through the trees is an engineering feat in terms of judging distances, weight, movement, and strain (see Chapter 6). In fact, learning to move independently through the trees (as opposed to just swinging on a branch, which they can do quite well even at the age of one year) is a long-drawn-out process and one in which the young orangutan may experience fear. Imitating precisely what mother does is one obvious way of learning to move independently through the trees. To avoid falling, orangutan juveniles at first keep one hand tightly attached to their mother's coat while following her. Later they may venture to follow her footsteps unaided.[20] It seems that imitation is involved.

A juvenile following the mother can be a comical sight. Most of us have seen clowns, as part of street theater, following unsuspecting people and copying their gait and limb movements. The crowd delights in this, especially if the clown is skillful. It is no less comical in orangutan displays. Davenport describes imitation by a juvenile who was being weaned. The juvenile followed the mother's every movement. It copied the movement of her arms and legs, stopped when she did, proceeded when she did, and negotiated the same branches.[21] The need to learn locomotion first by imitation may be dictated by the danger involved.

Later, the orangutan has to make more difficult judgments to cross gaps in the canopy. Bridging gaps by swinging on flexible trees or branches is, perhaps, the most difficult form of locomotion for an orangutan to learn; this behavior is not used for independent locomotion by infants until they are over four years old.[22]

The old divide of instinct versus learning[23] has not entered the orangutan climbing debate for the simple reason that humans have now raised so many orangutan infants that we know they cannot climb naturally. We mentioned earlier that orangutan babies separated from their mothers have to be taught to climb (see Chapter 3). They need daily lessons in it, starting with ropes and nets no more than a meter above ground, similar to the equipment used in playgrounds. At Sepilok Rehabilitation Centre, such classes are run on a daily basis. The kindergarten activities are a particular delight to watch. Two-year-olds are taken to the site, some walking and some being dragged along or carried and placed on the ropes. The infants usually know their routine: They love swinging on the ropes and seem to need no encouragement to engage in this activity, but they need a trainer to actually make them climb. The trainer is also there as assurance in case of a mishap and as a confidence builder. While the grip of orangutan hands and feet is usually very strong (when they are healthy), the coordination needed for climbing is difficult (taking one hand off the rope and moving it to another position while simultaneously doing the same with the feet).

Those orangutans who graduate from the rope nets are then asked not just to propel themselves forward but upward. Unless orangutans are trained to move about high in a tree from early on, they are more often than not very uncomfortable with height and will descend a tree as soon as allowed.[24] For an arboreal species, fear of height is surprising, but fear and caution in themselves are qualities that have survival value in a forest canopy.

Learning to Feed

Imitative learning plays a special role in feeding. Orangutans are omnivorous to some extent (see Chapter 3). They forage mainly for fruit and leaves, but they also eat nuts, insects, bulbs, eggs, and even bark and other vertebrates such as small birds, mice, and lizards. Altogether, more than 130 plants have been observed to be part of the staple diet of orangutans, and when invertebrates and vertebrates are added, some estimates suggest there may well be over 400 items on their menu. Such a variety of food is associated with availability (see Chapter 10), but how do orangutans learn to distinguish what is edible from what is inedible? How do they obtain the foods and process them?

The answers are probably more complex than we are able to untangle at the moment. There are now some studies of free-ranging orangutans indicating that imitation learning may be involved in acquiring these skills. Russon and Galdikas found, however, that age and social position influence motivational levels, and these in turn may influence the kind of learning that proceeds.[25] A young orangutan, for instance, may learn largely from the mother or a caregiver because of the attachment it has to that parenting figure, whereas an adolescent may spend more time imitating its peers. Very assertive caregivers may also find that more of their behavior is taken as a model than that of less assertive caregivers.[26]

They may also learn from experimentation, although this tends to be done very cautiously. For instance, we saw an adolescent orangutan who had been given a berry. The berry came from the market, and it was possible that it did not grow in her patch of the forest. She took the berry and placed it carefully on her lower lip, where it stayed for a considerable time while she inspected it by protruding the lip and looking downward. Then it was slowly moved about by the lower and upper lips, disappearing into the mouth only to be catapulted out to the same position on the lower lip as before. Then the berry was relocated back in the mouth and squeezed ever so slightly, followed by reinspection on the lower lip. The properties of the berry were thus gradually being examined: It did not move, it did not sting, it did not smell or taste unpleasant. Shortly thereafter the berry was consumed. Presumably, the next exposure to such a berry would invite a faster response, although this depends on the state of the original learning and the temperament of the individual.

So far, we do not know whether the level of difficulty is associated with imitative learning or whether some active teaching may be essential. There has been a sighting of active teaching in chimpanzees: A mother or close relative was observed to show the infant how nuts can be cracked by using a stone as anvil and hammer.[27] There are no observations, to our knowledge, of active teaching by an orangutan mother.

Active teaching may involve showing a task several times over, placing a tool in the hands of a juvenile, and guiding its way or teaching by reward. It would be feasible to think that such active teaching exists, considering the range of difficulties in preparing a food for consumption, but it is notable that it has not been observed.

Food extraction and processing require various skills and maturation levels, as well as appropriate assessment before attempting to obtain a specific food. Some fruit may be sweet to eat but dangerous to gather because it grows at the end of slender branches and requires skill to obtain it without falling. There are other fruits that are safer for orangutans to obtain—the fruits of what the botanists call "primitive" trees.[28] Durian, *langsat*, and jackfruit belong in this category because they bear fruit on the trunk itself or on the lower, thicker stems, and the fruit is thus placed "within easy reach of the larger animals who will eat it and disperse the seed."[29] Some fruits need peeling, others breaking. Such acts require motor coordination and specific skills. It may be easy to eat a fig, but it is not easy to gain access to the fleshy parts of a durian, for instance (see Chapter 3). Young orangutans tend to have difficulty achieving this feat and usually try to get easy pickings by sneaking pieces from their mother. However, orangutan mothers tend to restrict, or even deny, access. Why orangutan mothers deny food to their offspring is not clear. It is possible that by an occasional sampling and then denial they raise the motivational levels of young orangutans to a high pitch and so promote learning of the task in this way. Food that has to be extracted from below the ground or from logs (extractive foraging) may be difficult to access. Ants and bees, for instance, may attack and bite, and special methods of extraction are needed. Extractive methods of foraging, which often involve tools, require an apprenticeship,[30] and such apprenticeship, at least in part, may be served by imitating the parent.

Taking Risks and Thinking

Learning must take place in other ways as well as through imitation. Deduction, trial and error, and self-experimentation are other methods of acquiring new skills. Primates have large brains, long life spans, and long birth intervals, so they have plenty of time to collect experiences, skills, and knowledge. As in human parenting, the adult can provide only a baseline of information, skills, and ideas for the offspring, and it then depends on many things, such as temperament and memory, how well the early learning is put to use in later life. Life is full of novel situations and new experiences. In the animal realm, survival depends on competent adjustment, the effective processing of information, and adequate responses. Risk taking is involved. When the young orangutan finally leaves the mother, it is equipped with an impressive set of competencies and memories, but these can only form the basis for future activity.

Orangutans get more time to learn and ponder than nearly all other primate species, because often, they only start to reproduce some seven years after the end of adolescence. This is such an enormous expansion of the life-cycle events compared with most species that researchers have wondered about the purpose of a long state of juvenile and young adult life.[31] Some suggest there is a need for these extended nonreproductive periods for learning to be completed. They argue that this is necessary because the primates' environment (i.e., the tropical rain forest) is a very complex and dangerous environment in which to live. Primates need to know how to deal effectively with the risks before reproductive demands are placed upon them.[32]

Many of the demands of the environment could be met by trial-and-error learning, but other forms of learning must also be involved (e.g., insight and creativity, as discussed in Chapter 6). This may be particularly true of exploratory behavior in relation to new food items. We have noticed on countless occasions that free-ranging orangutans engage in long-drawn-out inspection of food before they consume it. They will inspect it by looking at the object in their hands, turning it and examining it closely. Sniffing new objects has been observed by Harrisson.[33] As we have shown elsewhere, levels of risk taking and exploratory behavior may depend on status, personality, and many other factors in primates.[34]

At Sepilok, infants frequently build nests for taking a siesta during the hottest part of the day. Once we filmed Bobbie absolutely absorbed in choosing suitably sized branches and inserting them, almost weaving them, into the nest as she sat on it. One particular branch was inserted several times before she was satisfied that it had been lodged in the right place. After an hour of concentrated work, the structure was still quite flimsy. Then she tested it to see whether it would hold her weight. Fortunately, she also took a firm grip on a branch with one hand, because the first time she slipped through. Thereafter, Bobbie did more testing and made many more corrections. Although the outcome was far from perfect and we felt very tempted to intervene and show her the "right" way of building a nest, the entire sequence showed that self-learning by trial and error might be happening. Much practice of this kind is likely to occur before the orangutan reaches adult body size and weight.

Beyond trial-and-error learning, there is something additional and very special about great apes. There is some evidence that they possess a representational map of the world. In other words, they can plan ahead and think. There are also a few, rare examples of orangutans being observed to do a little more than imitate blindly. For instance, the ability to cross a river by boat and even to use a paddle or pole for forward motion may still be regarded as simple imitative behavior. One orangutan at Tanjung Putting, however, may have demonstrated more than just that. He untied a rope that was fastened at the other end to the bow of a boat, took the rope in his mouth, got into the boat, and crossed the stream. On the other side, he alighted from the boat and went to the bank, dragging the boat behind him by the rope. He then held the rope in his hand or foot as he fed, making sure it stayed there for the return trip.[35] Although this may seem a simple example, the thought processes involved are complex. The orangutan understood that the boat would drift downstream if he did not keep hold of the rope. He also knew that he needed the boat to return to the other side—this shows that he had planned ahead. Such intentionality indicates a higher form of consciousness.

As we show in the next chapter, the ability to plan ahead and also deduce something from a novel situation and reset it into the context of the familiar is a very important ability that depends partly on the conceptual richness of the early learning environment and an ability that we (humans) have called intelligence.

6
PROBLEM SOLVING AND TOOL USE

Primatologists tell the following story about differences in problem solving between the great apes. A chimpanzee in a cage was given a screwdriver. He threw it away. A gorilla in a cage was given a screwdriver and he used it to scratch himself. An orangutan in a cage was given a screwdriver. He waited until the zookeeper had gone then used the screwdriver to unscrew the door and escape. This story is based on a real incident.[1] A young male orangutan had been captured and put into a traveling cage. From there, he had watched workmen nearby. When the men went home, they accidentally left a screwdriver within reach of the orangutan. He pulled it into the cage and proceeded to unscrew the bars. He would have succeeded in escaping had his action not been discovered. Since the orangutan was wild-caught, he would not have been exposed to screwdrivers before then and could only have learned how to use the tool by observing the workmen.

Orangutans are known to those who are close to them for their contemplative patience. Their powers of observation are probably unmatched, and they are also insatiable in their desire to manipulate objects. These qualities lead to an impressive ability to learn and solve problems. In the past, it was thought that the problem-solving abilities of orangutans were inferior to those of the other great apes, but this is not so. The reason for thinking this was due to the quieter, more contemplative personalities of orangutans.[2] They are less demonstratively interactive and often seem to be indifferent to what is being asked of them. This is particularly so of orangutans in captivity.

As we discussed in Chapter 5, when in captivity or rehabilitation centers, orangutans readily learn to imitate tool use by humans. They are also extremely resourceful in solving problems.

Problem Solving

In the 1980s, the comparative psychologist Jürgen Lethmate gave a number of problems to three young male orangutans to solve.[3] One task required the orangutan to use a series of keys to unlock boxes, in order to obtain a favorite sweet. There were five boxes, four of which could be opened with keys. One of the four locked boxes contained the sweet, another was empty, and the other two contained keys. Only one of these keys would open the box with the sweet. The fifth box could be opened without a key and contained the keys to the other boxes. The orangutan had to begin by choosing from this fifth box the key to open the box containing the key that would open the box with the sweet. Are you lost? Lethmate found that the orangutan was able to solve this problem without any difficulty. We conclude that orangutans might solve some problems better than we do! Orangutans enjoy fiddling with locks and bolts, and they will persevere at this longer than other apes.[4] They will even undo locks and solve problems without being rewarded with a sweet at the end. Just the act of manipulation and solving the problem is sufficient to keep their attention.

In another test, orangutans were tempted by some bananas hanging from the top of the cage, well out of reach. Scattered in the cage were several boxes of different sizes that could be stacked on top of each other, enabling the bananas to be reached. The orangutans, each tested separately, had no difficulty solving this problem. They simply stacked the boxes in the right order to make a stable pile and climbed them. There were one or two mistakes—two boxes were placed the wrong way round—but these were soon corrected. It was clear the orangutans did not achieve the solution to the problem by trial and error (i.e., by chance); they used reasoning to solve it.

One of the highest forms of problem solving involves insight. We have all experienced insight learning: We think about a problem and suddenly a way of solving it comes to mind and we feel a flash of inspiration. It is difficult to prove beyond doubt that an animal has solved a problem by "putting two and two together," but there can be signs that it has occurred. If an animal sits back, appears to be attending to the problem and then, more or less, goes deliberately to the solution of the problem, we might be correct in thinking that it has solved the problem by some kind of insight. If, on the other hand, the animal keeps trying different solutions to the problem, it may solve it eventually by trial and error. This takes a lot longer and does not require the same use of intelligence.

Out of thousands of trials that Lethmate gave the orangutans, he was convinced that insight had been used to reach the solution in only nine cases. One of these involved retrieving a sweet (a treat) to eat. A young orangutan was given a length of wood that could be used to poke the sweet out of a transparent plastic tube. He could see the sweet inside the tube. At first he bit the length of wood and then tried to insert it into the tube from the side and at angles—this was not successful. Becoming frustrated, he went away and sat down and began to perform repetitious behaviors with the piece of wood and his sleeping blanket. From time to time he glanced back at the tube with the sweet inside. Then, in a flash, he thought of the solution. He got up, walked over to the tube, inserted the length of wood into it and obtained the sweet.

Tool Use

The test in which the orangutan used a length of wood to poke the sweet out of the tube was an example of tool use, the length of wood being the tool. There are other examples of orangutans using tools in captivity. They have no difficulty using a long stick to knock down bananas suspended from the top of the cage. In a more recent study, an orangutan was tested, as were some chimpanzees and capuchin monkeys, with the tube containing a sweet and were given a variety of tools that could be used to obtain the sweet. They were given a single stick to poke the sweet out of the tube, a bundle of straight sticks that had to be separated before one could be used, or an H-shaped stick that had to be modified before it could be used effectively.[5] All the individuals tested could perform the simple task of using the single straight stick to obtain the sweet reward, as Lethmate had found before. The other tasks requiring modification of the tool distinguished between the capuchins and the apes. The capuchins did not untie the bundle of sticks and attempted, in vain, to insert the whole bundle into the tube. Nor did they modify the misshapen stick and so failed to insert it into the tube. All the apes, including the orangutan, untied the bundle of sticks to extract a single stick and successfully inserted it into the tube. They also modified the misshapen stick so that it could be inserted into the tube. The orangutan performed as well as the other apes. Perhaps this does show superior intelligence in apes compared with capuchin monkeys, as the researchers conclude, but later we explain that intelligence is related to lifestyle. There may be problems requiring tools on which capuchins perform better than apes.

As we have mentioned before, sticks are used as tools in the wild. Wild orangutans, as well as those rehabilitating, frequently break off small branches with the leaves attached and use them as swatters to keep mosquitoes away. In fact, they carry, use, and play with sticks much of the time. Sticks may also be used as tools to push or poke something unwanted away. For example, orangutans have been seen using a stick to knock a tree snake off a branch so that it falls to the ground. Broken branches are also used to scare away predators. Frequently, they throw them at approaching humans.

Making a tool is more complex than simply using one. Orangutans have been observed to make tools in captivity and in the wild. Captive orangutans will put together several short lengths of tubing, inserting the end of one into the next, to make a pole long enough to reach bananas suspended from the roof of the cage.[6] In the wild, they manufacture tools that are used for probing into holes in trees, presumably to obtain honey or insects such as termites, ants, or bees. This behavior has been seen in the swamp forest region of the Gunung Leuser National Park.[7] Orangutans here choose a stick and strip off its leaves. They then chew one end so that it becomes like a coarse paintbrush and split the other end so that it becomes a spatula. The spatulate end is held by the orangutan in its mouth and the brush end inserted into the hole in the tree. Then pounding begins by thrusting the head back and forth. Finally, the tool is removed from the hole, and the orangutan puts the brush end in its mouth and eats the insects or honey obtained from the hole.

In the same swamp forest, wild orangutans use smaller sticks, which they strip of their bark, to prepare ripe *Neesia* fruits for eating.[8] These are large fruits that split partly open when ripe. The seeds inside are eaten, but they are difficult to obtain because they are embedded in a mass of hairs that cause irritation. An orangutan holds the stick in its mouth, inserts it in the opened crack, scrapes out the hairs and removes the stick. The hairs are then removed from the stick by blowing or by wiping them off with the back of a fingernail. Next the seeds are pushed out with the stick and then eaten.

Strong sticks are used to break open termite nests for feeding. This form of tool use requires practice to reach perfection. Anne Russon observed the following incident at Sungai Wain Forest. Tono had been newly released there with some other young orangutans. He had been slow to learn how to obtain foods in the forest and had failed his class in breaking open termite nests. He was convinced that he should hammer them against a hard surface to open them, although that technique was not successful. The nest must be cracked open by hitting it with a large stick. Tono did not give up. He kept hammering away and could often be located in the forest by the sounds of his useless hammering. Once he was seen holding the nest in his fist and tapping it against his other hand while he seemed to be looking around for a hard surface on which to hammer it. He caught sight of another orangutan nearby and thwacked the nest on her head. She was both surprised and furious. She grabbed the nest from Tono's hand and threw it away. Tono had tackled the problem of opening the termite nest by various methods, the least successful being his final choice. He had failed to use an appropriate tool to solve the problem.

Orangutans use leaves as tools in a number of different ways. They may use them as a sponge to obtain water from holes in the forks of trees or to wipe unwanted substances from their coats. John MacKinnon observed a mother wiping her infant's feces from her coat.[9] Orangutans have also been seen wiping their faces with crumpled leaves.[10] These forms of tool use have all been observed in wild orangutans. We saw an adaptation of the use of leaves as a sponge for drinking in a rehabilitated orangutan. She used a stick of porous sugar cane to soak up water and then sucked on it to drink the water. She did this repeatedly.

On one of our visits to Sabah, we discovered a new form of tool use in orangutans using leaves as a plate from which to feed.[11] We first saw this behavior being performed by a rehabilitated adult female called

Gandipot. She had an infant of about two years of age at her side. Her life was now spent as a wild orangutan in the forest, but she occasionally came to a remote feeding table to take a snack of bananas. When we observed her, she had just taken a large bunch of bananas and moved a short distance away to peel them, which orangutans usually do by biting off the outside end of the banana and then squeezing the flesh into the mouth. Once her mouth was completely stuffed full of banana, she moved much further away, stopping on her way to pick a handful of large oblong leaves from a mango tree. Then she climbed high up in a sturdy tree and settled herself on a horizontal branch. We could see her clearly only through our binoculars and camera lens and, to our excitement, we managed to film the behavior that followed. It was the first record of using leaves as a plate.

Gandipot fashioned the leaves into a fan shape, holding the stalks together in one hand and spreading out the other ends of the leaves. Then she spat out the now mashed banana flesh onto the plate of leaves, examined it, and took it into her mouth again. This procedure was repeated several times before she consumed the entire amount. In the meantime, her infant tried to obtain some of the banana. At first he leaned over his mother's arm and managed to take a small amount with his lips before she pushed him away. His second attempt was rewarded by Gandipot holding out the food on the plate so that he could sample a little more banana. When she felt that he had had enough, she touched him on his back and he stopped feeding. Infant orangutans are known to beg for food from their mothers.[12]

This use of leaves was a chance observation the first time, but from then on, we kept our eyes open in the hope of seeing it again. It was not an easy activity to spot because it occurred when the orangutan was perched high overhead, but we were privileged to see the same tool-using behavior performed twice again, by two different females. In both cases the leaves were used as a plate in the same way, and banana collected from the feeding table was disgorged onto the plate before being consumed. These observations indicate that it may be a tradition among the group of orangutans that we observed. It could have been acquired as a result of observing humans eating from plates, but since the humans whom they see do not make plates from leaves, the orangutans must have invented at least that part of the operation. We think they could have invented the entire behavior themselves. It is also possible that this form of tool use might occur among wild orangutans. Since we found it difficult enough to see in rehabilitated animals, it would be extremely unlikely that anyone would detect it easily in wild orangutans.

Manipulation

Orangutans excel in manipulation, an ability they use constantly in the wild. Most of their day in the natural environment is spent manipulating vines and branches or fruits. They use manipulation to obtain food, make nests, and move around in the canopy. Every moment of the time they spend negotiating movement from one place to another requires manipulation and concentrated attention. Using the hands to do things other than hold on is risky when high above ground, particularly for a big animal. It is hardly surprising then that orangutans must concentrate on what they are doing with their hands and must develop excellence in manipulation.

Having such superior manipulation abilities requires a complex brain to control the hands and assess the information required for their use. That factor alone may be important in evolving a large brain overall, as Barbara Finlay and Richard Darlington reason in a paper published in 1995. These researchers examined how brains evolve by comparing the sizes of different regions of the brain in different species. They found evidence that if one part of the brain increases in size so that it is able to control a particular specialized structure (e.g., the hands), then most other regions of the brain grow larger, too. This may come about because the brain develops in a series of steps, and if one part of the brain needs to be larger, all the parts that develop with it and also those that develop after it must increase in size also.[13] It means that acquiring one highly specialized ability may increase the brain's capacity to carry out many other functions. Thus, using the hands to manipulate might lead to a larger brain and enhanced abilities to learn and solve problems.

Engineers or Trapeze Artists in the Trees?

Being such large animals, it is no small feat for orangutans to travel from tree to tree. A fall would be fatal or, at the very least, the cause of serious injury. Heavy bodies do themselves a lot of damage when they fall. Every move must be negotiated with extreme care—in fact, every move is a problem to be solved. Knowing not to grab hold of a branch that will not support the full body weight requires considerable skill. Some branches are thin but strong, while others are thick but brittle. Some vines will break under the weight of an orangutan, others will not. There are many different kinds of trees in the rain forest, some fast growing and weak and others very slow growing and strong. Orangutans need to be the botanists of the rain forest, able to distinguish the quality of trees.

Smaller primates, such as the long-tailed and short-tailed macaques that share the forest with the orangutans, must be skilled climbers too, but being much lighter in weight, they have a greater choice of branches to support them. They do not need to make the same distinctions and choices as the orangutans.

Small primates have far less trouble than large ones in moving from one tree to the next because many of the small branches on the outermost extremities of the tree can still support their weight. They can go right to the outer branches of one tree and on to the outer branches of the next, like tightrope walkers. If orangutans tried this, the branches would bend down and might break. They have to be trapeze artists rather than tightrope walkers.

The trapeze skills of orangutans are most impressive when they use flexible saplings to cross gaps in the canopy. They negotiate the gap by swinging back and forth in larger and larger arcs until they manage to bridge the gap and take hold of the next tree, while letting go of their swing at the same time (Figure 6.1). Sometimes they swing on liana vines to cross gaps, and they also use the vines just for playing. Not all locomotion by orangutans involves swinging. They walk on the ground sometimes and, more often, clamber about grasping trunks and branches with hands and feet going in all directions.[14] The term "clamber" has been used to describe this way of climbing, but orangutans are no less agile in this form of movement than in the other movements they use.

Every mode of locomotion that the orangutan adopts in the complex structure of the rain forest requires the solving of problems. And survival depends on their ability to solve them successfully. Are orangutans engineers or trapeze artists? Do they make detailed, even conscious, calculations each time they negotiate moving from one place to

FIGURE 6.1 *This infant orangutan was raised by humans and then confiscated. Here he is attempting to negotiate branches and trunks no higher than a meter off the ground. Without guidance, this infant will not become a competent climber.*

another, or does practice allow them to do it automatically? No one knows whether orangutans use their higher forms of thinking each time they move. Perhaps they do this at first, when they are young and learning, and then once the skills have been learned, they may be able to move around without having to think about it so much. They may begin independent life as engineers and graduate to trapeze artists. On the other hand, they may never be able to get around entirely without solving problems and using the higher processes of thought, even though they do learn their skills when young.

According to Daniel Povinelli and John Cant, the question of whether higher thinking (i.e., problem solving) is needed to move around in the forest depends on body weight.[15] They see this as the main difference between the great apes and monkeys, one that led to the evolution of higher intelligence in apes. Because monkeys weigh much less than the great apes, they can learn fixed ways of moving about in the forest, and once they have done so, they need to use very little brain power for the activity. The great apes may be too big to do this and may always have to think carefully about their actions. Their limb movements vary with each particular problem of locomotion. They have to be much more flexible than monkeys in the way they go about moving, and according to Povinelli and Cant, they have to "produce creative, on-the-spot solutions to immediate problems."

To be successful, they would need to plan what they are going to do and predict the likely outcome if they put their weight on this branch or swing on that liana vine. In other words, in their heads they would need to run through the possible outcomes of their actions to avoid taking risks.

Galdikas has suggested that it is because wild orangutans have to use up so much of their thinking on solving the problems of locomotion that they use fewer tools than orangutans at rehabilitation centers.[16] Although interesting, this idea implies that there is a limit to the thinking capacity of an orangutan's brain that once the brain is taken up with planning strategies for locomotion, there is no time or brainpower left for thinking about using tools. We think this is unlikely, and in fact, the opposite is more likely to be the case. Once a brain has evolved the capacity to plan ahead and solve the complex problems of moving a heavy body through the trees, it might also have acquired the capacity to plan other actions. Povinelli and Cant suggest that the requirement to solve the problems of moving has increased orangutans' self-awareness (i.e., a concept of self). We discuss self-awareness in the next section. Here, we want only to point out the effects that being a heavy animal living in the trees may have on the evolution of the brain and on the kinds of thinking that can be carried out.

Nest building is another engineering feat carried out by orangutans. Some researchers regard it as a form of tool use. It is very difficult to build a nest from branches and leaves while perched many meters above the ground, and to build one of sufficient strength to support an orangutan's body weight is especially demanding. Thus, it is possible that a need to solve "on-the-spot" problems in nest building, as well as in negotiating movement, has led to an increase in the orangutan's mental capacity.

Self-Awareness

To be aware of one's self is an aspect of consciousness that requires higher thinking. There are many ways of testing animals to see whether or not they have consciousness but little agreement on which tests really prove the case. Each test gives an indication but does not provide conclusive results. We do not intend to discuss all the details of testing for self-awareness in primates,[17] but we cover what little is known about self-awareness in orangutans.

When an orangutan looks into a mirror, does it know it is looking at itself, or does it think that the image is some other orangutan not seen before? Looking at a reflection in a human-made mirror, rather than a pond, is not something that an orangutan, or any other animal, does in the wild; so to test them with mirrors is relevant only to animals in captivity. Mirror-recognition tests are also relevant only to animals that have had some experience of looking in mirrors.

There have been very few studies of how orangutans behave toward their own images in mirrors. Most of the research on self-awareness in apes has been conducted using chimpanzees and, to a lesser extent, gorillas. Both species seem to recognize that the image in the mirror is of "self" and not "other." The few studies of mirror-looking in orangutans indicate that they too can recognize their own images.[18] If an ape recognizes itself in a mirror, we might expect it to use the mirror to look at parts of its body that cannot be seen otherwise. It has been found that chimpanzees use the mirror to look at their tongues and teeth, to observe while cleaning their noses, or to look at their anal and genital regions.[19]

In a recent study conducted by Ethel Tobach of the American Museum of Natural History and colleagues, a group of six captive orangutans was tested by hanging on the walls of their enclosure a mirror and life-size photographs of each individual in the group.[20] They all behaved differently toward the additions to their environment, but most used the mirror to look at their genitals and other parts of their bodies. This behavior suggests that they recognized themselves in the mirror. One adult female spent much of her time looking in the mirror, and she also looked for longer times at her own portrait than at the portraits of the other orangutans in her group. This result shows that she could distinguish between portraits of orangutans familiar to her and the one (of herself) not familiar to her. Whether she saw her own portrait as self or as the image of some other unknown orangutan cannot be decided. Only the behaviors performed in front of the mirror (i.e., looking at parts of her body that could not be seen otherwise) tell us that she might have been aware of the fact that she was looking at herself.

Povinelli and Cant, as discussed above, hypothesize that the orangutan's heavy weight and mode of locomotion have led to the evolution of self-awareness. They also suggest that the life in the trees of the ancestors of all apes, including humans, was an important factor in the evolution of increased intelligence.[21] If their hypothesis is correct, small monkeys that do not have to use so much mental capacity to negotiate moving in the trees should lack self-awareness. They should not, for example, recognize themselves in mirrors, but there is at least some evidence that they do. Marc Hauser and colleagues at Harvard University tested some small South American monkeys, cottontop tamarins, with mirrors.[22] Some of the monkeys used the mirror to look at parts of their bodies they could not otherwise see, just as orangutans and other apes do. Also, marking the monkey's white piece of hair on top of the head with colored dye caused them to touch this part of their bodies more than usual. They seemed to recognize that their hair had been changed and that the change was to themselves, not the image in the mirror. If they had thought the latter, they would have touched the image in the mirror instead of themselves.

This result indicates that a species of small monkey may have self-awareness and does not support the hypothesis of Povinelli and Cant (i.e., that being large primates living in trees leads to higher intelligence and to self-awareness in apes). There are many ideas about the mental capacity of apes and how or why it might have evolved, but, so far, there is no strong evidence for any of these ideas. We think that there might be no single aspect of life that explains the evolution of superior abilities in thinking and solving problems.

We also question what is meant by "superior intelligence." The great apes are close to us on the evolutionary tree and so would be expected to share with us more of the same ways of thinking than other animals do. Yet birds can solve very complex problems, as well as primates in some cases.[23] Each species has its particular abilities that are applied to special problems relevant to survival in its own place in the scheme of life. What we call "intelligence" is often a narrow view and reflects similarity to humans. There is probably no unitary concept of intelligence that can be applied inflexibly to all species, and it would seem irrelevant, if not impossible, to rank the intelligence of animals on a single scale based on their abilities to perform a single set of problems. For the same reasons, we reject comparisons that attempt to rank the intelligence of the different species of apes. Chimpanzees are not necessarily more intelligent than orangutans or gorillas. It depends on what problem they have to solve and in what context they are tested.

Sharing Problem-Solving Skills

Our earlier observation of three orangutans at one site using leaves as a plate for feeding prompted the suggestion that this could be a tradition in that group of orangutans. We thought that this form of tool use might be handed on from one individual to another and could even become a cultural tradition that is passed from one generation to the next. The same may be true of the tool use observed by van Schaik and Fox in the population of orangutans living in the swamp forest region of the Gunung Leuser National Park (i.e., making and using a spatula to probe for insects or honey and using a stick to eat *Neesia* fruits). We have no evidence that this is the case, but it is significant that such tool use is demonstrated by more than one individual in the same area. It raises the possibility that one individual might learn from another.

One orangutan could learn by watching another make and use a tool. This would be learning by imitation (see Chapter 5). The other alternatives are that one individual teaches another or that each individual acquires the skill quite independently. So far, there has been no observation of one orangutan actively teaching another. Imitation learning is more likely than direct teaching.

In Chapter 5, we gave examples of orangutans imitating the tool-using skills of humans working at rehabilitation centers. Josep Call and Michael Tomasello of Emory University question whether orangutans really acquire new skills by imitating.[24] They think that the orangutans may simply mimic the behavior of humans without understanding what they are doing and thus without being able to apply the learning to solving a problem. They base this conclusion on tests given to fourteen juvenile and adult orangutans in the laboratory.

First, the orangutans had to watch either a human or an orangutan solving a problem to get a sweet reward. The reward was inside a box constructed in such a way that it could be opened only by manipulating a rod, projecting from one of its sides. The rod had to be manipulated in a specific way in order to open the box and obtain the sweet reward. After watching this task being solved, the orangutans were given it to solve themselves. They showed no signs of having learned by imitation. This result led the researchers to conclude that "apes do not understand the goals (i.e., intentions) of others, at least not in the sense of differentiating means from goals." They also concluded that the orangutan's "social learning skills are of limited usefulness in problem-solving contexts." In fact, they decided that this is true of all apes. To state their conclusion simply, apes may mimic the behavior of another, but they do not understand what they are doing.

We think this is too sweeping a conclusion to make from just one experiment testing a few subjects on only one problem. Another kind of problem in another situation might have produced a very different result. Also, the orangutans used in the experiment were in captivity, which gives them a much more limited life experience than either rehabilitating or wild orangutans. They were institutionalized, and that may have reduced their interest in the task they were given.

Motivation for Problem Solving

Even the details of life in captivity can be very important. For example, being raised in a nursery by humans as compared with being raised in captivity by another orangutan may have profound effects on the orangutan's behavior in general and also in specific tests. Call and Tomasello recognized this in a paper they published a year before the one they wrote on imitating in orangutans.[25] They found clear differences between orangutans in their ability to understand the meaning of pointing at an object. An orangutan called Chantek, raised in close contact with humans and taught sign language, displayed much better understanding of the meaning of pointing to show a human where something is hidden than did an orangutan raised in a less intense relationship to humans. Chantek would point to show a human experimenter the location of a tool that was needed to open a container of food. He was willing to assist the human in solving a problem (finding the tool). The other orangutan would not do this. Either she did not understand the meaning of pointing or she was not amenable to taking part in the experiment.

The fact that an orangutan, or any other animal, fails to solve a specific problem set by the experimenter does not necessarily mean that the animal is incapable of solving the problem. Lack of motivation to do what the experimenter wants may be the explanation for not solving the problems presented in these experiments. There is also the possibility that the orangutans deliberately fail to solve the problem. They are capable of knowing what is wanted and choosing to do the opposite.

We believe that the background of every orangutan has to be taken into account when interpreting the results of laboratory experiments that test problem solving and learning. Where possible, the same approach should be taken when interpreting observations made in the wild. General statements about what can be done by orangutans as a species, or apes as a whole, should be made with caution.

When Chantek pointed out the location of the hidden tool to the human experimenter, he was communicating. The reverse situation can also occur. An orangutan is able to understand pointing by a human, as shown by Shoji Itakura of Emory University and Masayuki Tanaka of the University of Kyoto. They tested chimpanzees, human infants, and an orangutan with a choice of containers, one of which contained a piece of food. The human experimenter attempted to help the subject to find the food-baited container by pointing to it. All three species were able to interpret the pointing correctly.[26] This result contrasted with a report a year earlier saying that apes, including orangutans, were unable to do this.[27] In fact, Itakura and Tanaka were able to show that the apes could understand "pointing" by direction of eye gaze only. If the human experimenter gazed closely at the container with the food instead of pointing with the arm and hand, the apes were able to make the correct choice. Understanding communication by direction of gaze is relevant to their own forms of communication (see Chapter 7).

Lastly, we discuss the ability of orangutans to solve a problem that human children, on average, are unable to do before the age of seven to eight years. The task requires the child to tell the difference between two volumes of liquid regardless of the shape of the containers. For example, the child might be asked to choose the larger of two volumes of juice (one volume being twice the other) presented in different-shaped transparent containers. This task is simple when both containers are the same, but it starts to get difficult when the larger volume is in a glass with a large diameter and the smaller volume is in a tube with a small diameter. Now the surface of the smaller volume is higher in its container and the young child usually chooses this one.

Josep Call and Philippe Rochat tested four orangutans, one of them Chantek, on tasks such as this. The orangutans had to point at the container with the larger volume, and their reward was getting it to drink. They had no difficulty being motivated to take the larger of the two volumes. They also had no difficulty in recognizing the larger volume irrespective of the shape of the container in which it was presented. The experimenters also tested ten human children aged six to eight years on exactly the same tasks. Most could perform the task, but three or four of them had more difficulty than the orangutans. In fact, two of the human children could not make the correct choice at all. Both the orangutans and children found it difficult or impossible to do the task when it was made more difficult by pouring the larger volume of liquid into a number of smaller glasses. It was more difficult to add up these separate volumes and see them, collectively, as being larger than a smaller volume in a single container.[28]

Later experiments by the same researchers showed that the orangutans were able to accurately assess and determine the volume of liquids in containers of different shapes, but they had some difficulty tracking the same liquid as it was poured from one container to another or from one container into a number of containers. However, Chantek was able to perform almost all the different tasks that the experimenters gave to him. Call and Rochat said that his superior performance might be the result of being raised in exceptionally close contact with humans (Chantek was trained to use sign language to communicate with humans; see Chapter 8).[29] In other words, they believe that he had acquired something exceptional for an orangutan by living and communicating with humans. This view reminds us of the past, when it was thought that "primitive" people gained, intellectually and culturally, by being adopted and raised by white colonial families. No one considered what they lost by being taken away from their own culture because it was thought not to count.

Likewise, all the orangutans raised and tested in captivity have lost their own culture. If Chantek performed well, it might have been that extra human nurturing went some of the way to replacing his own losses and was better than the more deprived conditions under which the other orangutans had been raised, even those who had stayed with their own mothers in captivity. We suggest that rather than acquiring something special from his human carers, Chantek managed to retain abilities that impoverished living had taken away from the other orangutans who took part in the experiment. In the final analysis, these experiments tell us more about the paucity of laboratory environments for raising and keeping orangutans than about the cognitive abilities of the species.

Plate I BJ was our constant companion at Sepilok.

Plate II Different faces and hairstyles of orangutans.

Top row: Merotai (left) has mottled skin—orangutans have color variations of skin like humans, including black, pink and an olive-colored complexions; Gino (middle), a very charming and pleasant three-year-old, has pink eyelids; Kurasi (right), a subadult male, with a very gentle facial expression.

Middle row: Alice (left) wears a punk-style hairdo; Clementine (middle), shown here during a very fleeting meeting—thirty seconds later she was gone with her offspring; Raja (right), who has very little hair, is no older than Kurasi but is larger and appears more authoritative and more intent on mischief—substantial individual differences among peers of the same age and sex.

Bottom row: Rampig (left), a very active young female, has a comical look and wispy hair shaped like two horns; Boon (middle), a young, serious, and rather anxious juvenile; Kurasi (right) wears his hair swept forward.

Plate III *Judy is almost hairless, and her pink and soft skin gives her a human appearance, especially when she copies human behavior by walking upright, taking buses, and sitting on park benches like a lady!*

Plate IV *Rampig uses her feet to break a twig.*

Plate V *Since the birth of her offspring, Jessica has changed. Her previous depression has lifted, and she now smiles most of the time.*

Plate VI *Clementine and her eighteen-month-old male offspring feed at the same time. Note how the infant mimics his mother's behavior in a mirror-image fashion.*

Plate VII *Gisela and Lesley meet Abbie for the first time. Abbie did not need any persuasion to alight the lap of her "new mother."*

Plate VIII *Two years after our first visit, we found Abbie again, and despite our absence, she showered Gisela with attention.*

Plate IX *Our third meeting with Abbie. Almost an adult now, she has taken on premature responsibilities as surrogate mother for the orphan Tom. During the time we observed them, Tom never allowed Abbie to be more than a meter away from him.*

Plate X *Gaze avoidance. Here, Puspa, a Sumatran female orangutan at Perth Zoo, avoids looking directly at Gisela (and the camera), glancing sideways at another orangutan instead.*

Plate XI
This Sumatran adult female orangutan, also at Perth Zoo, looks sideways at Gisela (and the camera). Direct gazing with face and eyes facing the onlooker is not common in orangutans.

Plate XII Hsing, an adult male Sumatran orangutan at Perth Zoo, looks down and exposes his silver-colored eyelids.

7 COMMUNICATION

Although very little has been written about communication in orangutans, few researchers would fail to comment immediately on having a sense of communication with them. This is so because many communication signals that orangutans make seem close to those that occur in human society. This fact brings with it the risk of misinterpreting what orangutans are attempting to communicate because interpretation of the meaning will be influenced by the observer's cultural background.

We recall an instance at a zoo where a group of people were fascinated by one particular orangutan because she had come over to them and was facing them directly, eyes fixed on their faces. She waved her hands toward them, repeating this several times. It was the same motion that orangutans use to throw sticks. One observer said how nice it was that she "wanted to speak" to them. The opposite was true: This signal in orangutan communication is a strong warning to go away and is not in the least friendly.

Another example is the baring of teeth, which we tend to read as a smile. In chimpanzee communication, as in that of many other primates, it is usually an expression of fear or aggression. This may well also be the case in orangutans. "Smiling" at a chimpanzee or an orangutan may therefore not be a friendly gesture to make, but we invariably do so because the smile has become a cultural human symbol for friendliness.

Social communication in orangutans is more difficult to study than in the other apes because wild orangutans are not seen in groups or pairs very often. In groups, we can assess whether an individual's gesture, vocalization, or eye expression has a communicative function by watching the responses that it causes in others. In the orangutan world, mother-infant pairs probably offer the most material for studying communication. Often, though, the human observer sees just one individual, and if there is any social interaction, it is with the observer. On the other hand, orangutans in rehabilitation centers can be very gregarious and provide many examples of communication,[1] as we discuss in Chapter 3. The same may be true in zoos.

Even though orangutans are semisolitary and do not display the same demonstrativeness as group-living species, they still have complex and socially intense communication. Communication in orangutans is achieved by vocalizations, gestures, and other visual signals. They also make use of smelling (olfaction) and touching. We believe that visual signaling is the most important.

Visual Signaling

It may be difficult to accept that not so very long ago, people believed that humans were superior to apes because only they (humans) were thought capable of expressing emotions.[2] We now know that orangutans express a range of strong emotions such as anger, fear, anxiety, pleasure, and love and that many of these emotions are displayed as visual signals on the face, in body posture, or in the movement of limbs. For instance, when orangutans are in great pain, they place their arms behind their heads and often close their eyes. When Abbie began to understand the impending separation from Gisela, her whole body posture changed and her face contorted in fear.

It is important to distinguish between intentional and unintentional communication. Most species, including humans, display unintentional signals that are communicated, without either animal or human being conscious of them. These unintentional signals betray states of emotion. Intentional signals are produced by a conscious decision to communicate a message. It is difficult to decide, however, whether an animal is communicating intentionally or unintentionally.[3] Probably most of the communication by orangutans discussed here is intentional, although this has not been proven. Few researchers have carried out controlled tests to see whether the communication is intentional or not.

Gestures of the Body

Communication by gestures has been a neglected field in studies of great apes. This is surprising because so much of their communication seems to be expressed in gestures and they are therefore likely to be important. The work that has been done on great ape gestures has largely been confined to chimpanzees and gorillas. Orangutans have been neglected, except for a handful of recent studies.[4]

Visual displays by orangutans may involve the whole body, including posture and movement. Stretching, jumping, or arching the back can communicate a range of different messages. Locomotion itself can be a form of communication. As humans, we tend to notice a variety of different gaits and corresponding body postures. Orangutans are not at their best when they walk. Their gait tends to look particularly awkward when they attempt to walk upright, and they generally do so only after much exposure to humans (e.g., Judy and BJ both walked upright). However, sometimes they display great bursts of running toward a goal (e.g., to get to a fruit tree, to ward off intruders, or, apparently, even to express joy). Willie Smits from the Wanariset Rehabilitation Station in Indonesia recounted an incident, which he interpreted as joy, at the time of releasing an adult orangutan. When the rehabilitation team took her out of the crate and released her in view of a lush patch of forest, she was seen running back to the forest with arms swinging wildly above her head, hands waving in quick motion.[5] This is not the way orangutans keep their balance: She was expressing emotion by arm and hand gestures.

Many of the limb and head movements made by orangutans remind us of human signals (i.e., waving, head shaking, moving, or raising arms or hands). They also have social signals that we, as a species, have largely lost, such as lip-smacking and hair-bristling. Orangutans use all these visual signals to great effect, not just to communicate with other orangutans but also occasionally to communicate with other species. We once saw Simbo interacting with a troupe of short-tailed macaques (Figure 7.1). He and the macaques had spotted food at the same time, and some of the macaque males got to it first. Simbo almost flew to the spot and, while still in full motion, swung his left arm around in a very powerful motion in the direction of the macaques. He also shook his entire body and emitted a low but audible grunt, all in one fast inte-

FIGURE 7.1 *Macaques are smart and quick monkeys who live in groups. However, a single male orangutan will send them scurrying for cover.*

grated action. We were glad we were not in his path. His actions showed that he was not one to meddle with.

More important, the macaques seemed to think so, too. They were screaming and running in all directions. Short-tailed macaques are usually an unafraid and rather audacious species, ready for a fight and prepared to get their share of anyone's table. As a group they can be quite dangerous—the males form a charging frontline that will pounce on intruders quite readily and then be joined by other adults in the group eager to enter the brawl. But after Simbo had arrived and claimed the food, they sat there meekly, a good distance away, and did not move. They did not make a sound or attempt to challenge Simbo while he was feeding. His actions were a final threat, and his message was clearly understood as a non-negotiable claim.

Among mammals we do not often find objects being used to enhance visual signaling, but orangutans and a few other primates use them. Orangutans use their limbs in very expressive ways, as MacKinnon found when he was tracking them. They made motions of waving their arms away from their bodies, signaling to him to leave.[6] When he failed to respond, they sent stronger messages that were unmistakably aggressive, first shaking branches and then fashioning them into harpoons to throw at him. Branch shaking has also been observed in gibbons and chimpanzees.[7]

Baboons throw stones at predators. A wild orangutan male that we saw thumped his arms and thrashed on his nest. This behavior seems to have the same communicative function in both Sumatran and Bornean orangutans, despite the fact that they have been separated from each other for thousands of years.

Gisela studied some captive Sumatran orangutans at Perth Zoo. One seven-year-old female by the name of Sekara always saved a stick especially for Gisela or, rather, for the camera and threw it when the time seemed auspicious. Karta, an adult female, generally bad-tempered and disinclined to be friendly toward anybody, regularly took a mouthful of water that she would squirt with obvious pleasure as Gisela walked past her cage. Karta knew the exact distance she was able to squirt the water and could judge when the human observer was too far away to hit. She would save her mouthful until Gisela stepped just within shooting distance and then release the copious volume of water from her mouth.

Whether arm movements or the throwing of water or sticks, the message is unmistakably an antisocial one.

One male orangutan, BJ (Plate I), used his limbs to send antisocial messages. He had learned that humans ran away when he charged at them with his arms over his head and hands forward, grimacing at the same time. It was an impressive display and it worked. Visitors to Sepilok always ran away as fast as they could. But the most surprising aspect of BJ's body language was his gesturing to humans for invitational purposes. A place that was off-limits for orangutans was the Research Centre, which was located just outside the reserve. The young orangutans knew full well that they were not to go there. While the keeper was busy talking to visitors one day, BJ came up to Lesley and took hold of her hand in an encouraging way. Usually, he was immediately ready to do something naughty once we were within reach of his arm, but he behaved charmingly on this occasion, his hand resting nicely in Lesley's with a sense of a shared conspiracy. He tugged her hand and gestured encouragingly with his head in the direction of the exit path from the reserve. He repeated this surreptitiously each time the keeper looked away. The aim was clearly to solicit help to leave the reserve and go to the center.

Age differences may have an influence on what and how something is communicated. For instance, it was found some years ago that chimpanzees have a special range of gestures used exclusively by infants and another range of gestures used only by adults.[8] Too little work has been done in this domain to confirm whether this is true for orangutans, but Kim Bard has studied some forms of communication in young free-ranging orangutans and showed that they are intentional.[9] She demonstrated, for instance, that orangutan infants and juveniles show gesturing to the mother to obtain food. Usually the infant holds a cupped hand, palm facing upward, underneath the mother's chin.[10] This may well be a gesture that is exclusive to infancy and "childhood." We have observed Abbie pinching the chin of another orangutan eating bananas, apparently with the aim of getting the recipient to give her some food. Chin pinching may well have developed from the cupped-hand-under-the-chin gesture.

The Face

Another specific region of the body from which signals can emanate is the face. Much research on communication by facial expression has been carried out on primates, and this is a large field of investigation, although little research has been done on orangutans.[11] The use of facial expressions by primates for communication has been of interest partly because a primate's face is similar in anatomy to the human face and partly because Darwin singled out the face as an important site for the expression of emotions.[12]

Orangutans seem to display many of the same facial expressions as other primates, including humans (Figure 7.2). For example, when playing, they show a relaxed open mouth, and when they are worried, they may bare their teeth in a display of submission.[13] When angry, they make a threatening expression by grimacing, baring the teeth, or opening the mouth widely.[14] According to MacKinnon, fear is expressed by drawing back the sides of the mouth, which also exposes the teeth. Mild anxiety is expressed by pouting of the lips, and a threat is accompanied by a "trumpet" mouth (shaping the lips into a trumpet; see photo at beginning of this chapter) and deep grunts followed by gulps.[15]

Early work by Jan van Hooff from the University of Utrecht, the Netherlands, has shown that human laughter and smiling may have analogues in primates.[16] There is a facial expression that we might call "smiling" that involves turning the corners of the mouth slightly

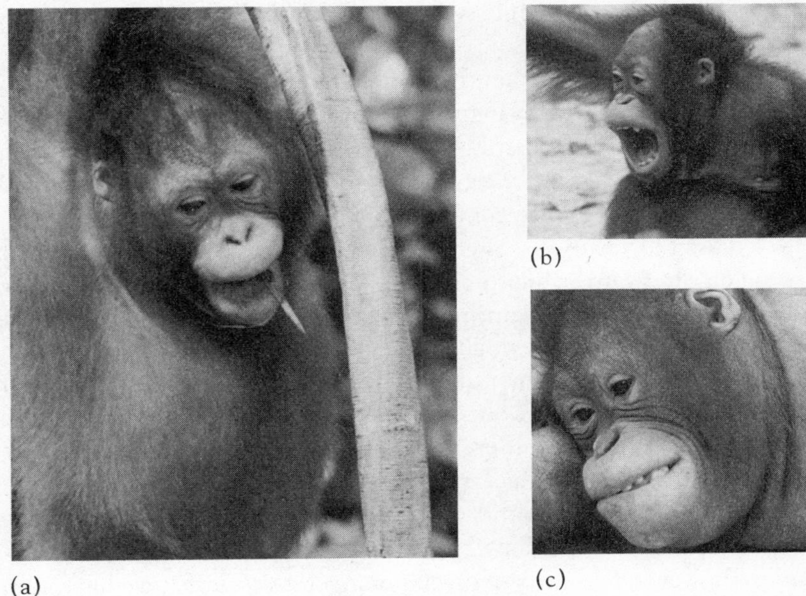

(a) (b) (c)

FIGURE 7.2 *Facial expressions. Playfaces made by the subadult Abbie (a) and by her playmate Alice (b). The meaning of the facial expression in (c) is unclear, but we observed this grimace quite often among rehabilitating orangutans.*

upward and exposing the teeth. This expression may be silent or accompanied by sound. It has been described as one of the oldest facial expressions in our evolutionary history because it has been found not only in primates and humans but in many other mammals as well.[17]

But it has probably undergone a change in meaning. In human communication, a smile is a friendly gesture, but the origin of the silent bared-teeth display is actually fear and has been associated with a threat or a strongly aversive stimulus. We described smiling in Jessica after the birth of her infant, where the mouth was completely closed and the corners of her mouth turned upward. The teeth were not exposed (Plate V). If not happiness, this closed-mouth display suggests contentment. Whether Jessica's signal was intentional is doubtful.

Van Hooff made the interesting suggestion that in terms of the evolution of facial expressions, smiling and laughter may have had little to do with each other. He thought that laughter and smiling had different roots. Laughter is associated with breathing and breathing technique (i.e., a vocal activity), while smiling is related only to movement of the facial muscles.

Another facial expression that is found in many species, as well as in humans, is the yawn. In mammals, yawning is probably more widespread than has been described in the literature. When humans yawn, it usually means that the yawning person is tired or bored, as a detailed study by Robert Provine confirmed.[18] Orangutans yawn (Figure 7.3). Jessica used to yawn very often prior to the arrival of her baby but not so much afterward, which might suggest she is no longer bored now that her baby occupies her time. We have no conclusive evidence, but yawning in orangutans might also mean something different from human yawning. In many primates, yawning may be an expression of embarrassment or frustration.

One of the notable differences between free-ranging and captive orangutans is their facial expressions. We have stood at orangutan enclosures around the world watching them and found relatively few facial expressions over long stretches of time. In free-ranging orangutans, however, in groups of several orangutans together, we have seen many different facial expressions. Unfortunately, sadness of the face and the eyes is the image that is now the most prevalent and most photographed expression of orangutans (Figure 7.4).

The Eyes

One of the remarkable features of great apes is their excellent eyesight and color vision. This has implications for communication. With excellent eyesight, visual signaling may take on a special role in communication. Finer and more complex signaling would be possible, and the eyes could become a focal point for exchanging messages. Some of these signals may be difficult to detect by the untrained human eye. Eye movement and eye contact among orangutans is poorly understood, but it now seems that eye movements could be a more important means of communication in orangutans than first thought.

The eyes can express fearfulness, anger, curiosity, and real or feigned indifference. Eye expression can also be used to control and facilitate actions. Several writers point out that the role of stares in primates is rather complex and may have different functions in different situations. For instance, the stare may be used as a threat or a reprimand by Old

FIGURE 7.3 *The yawn of a male orangutan is awe-inspiring. It is not clear whether the gesture is used intentially to intimidate or whether it is merely an expression of boredom or frustration.*

World monkeys.[19] A male gorilla uses a stare in conflict resolution if grunts have been unsuccessful in settling squabbles between females or juveniles.[20] The stare may also be used in bonding. Bonobos and chimpanzees use mutual staring as a form of positive contact. In bonobos, a direct stare may be a way of seeking a sexual encounter, or it may be a reprimand or assertion of dominance.

We have seen only one stare in an orangutan. A keeper had removed her food from the provisioning site in Sepilok and she stared at him. In fact, she looked very angry but did not take any action. The stare was lost on the keeper as he kept his back to her. Galdikas has reported

FIGURE 7.4 *A sad expression is on the face of this orphan orangutan, Gemasil. He was too young to be left alone and died shortly after this picture was taken.*

direct staring during an encounter of two adult male orangutans, followed by a brief and hostile encounter.

Staring in primates is often preceded by lowering of the eyelids. The eyelids then become exposed, and in orangutans and other primates, they can be rather spectacular. In macaques, the eyelids are white and in some baboon species a silvery color. In a threat display, a male baboon flashes his colored eyelids, making the signal more threatening, or at least more conspicuous. Raising the eyebrows, either by retraction of the scalp or independent eyebrow movement, also reveals the eyelids.

As we found in our own study of orangutan eye movements, eyelids can mislead the onlooker about the direction of gaze. When orangutans lower their eyelids, they reveal a small silver lining at the base of the upper eyelid (Plate XII). This makes it look from a distance as if the eye is turned in one specific direction (i.e., as if the individual is watching something intently). It is not clear whether the orangutan knows of its own capacity to deceive. The silver lining on the upper eyelid seems to be more pronounced in the Sumatran orangutan than in the Bornean orangutan.

Bard has observed that orangutan juveniles beg for food from the mother by shifting their gaze back and forth between the mother's eyes and the food item, as also occurs in chimpanzees.[21] The cue is to lead the mother to follow the infant's eye gaze to the desired object, and the intention seems to be to communicate that this is an item the juvenile wants. This requires the mother to associate the eye movement with the desire of her offspring (i.e., to understand its state of mind).

Chantek (discussed in Chapter 6) was found to alternate his gaze between a human's eyes and the point he wished that person to look at.[22] The evidence for this ability to follow the direction of gaze and to use eye gazing as a signal for someone else to attend to an object is mounting all the time. In terms of cognition, this is a tremendously important and impressive step. It indicates intentional communication and consciousness. To follow the direction of eye gazing of another individual requires an understanding that the other is looking at something different.[23] Presumably, orangutan juveniles would not employ this technique if it were outside their conceptual understanding and if it never led to a successful response.

Eye movement and attention can also express social status and social hierarchy in what is generally referred to as the "attention structure" among monkeys and apes.[24] Those in lower-ranking positions in the group do more looking toward higher-ranking individuals than vice versa. Communication channels are thus not evenly distributed. In chimpanzee society, orders may come from higher up, from a dominant male, which the junior has to follow, and only rarely will the junior have an opportunity to say something in return that will in any way require a response from the higher-ranking individual.

Our own study into the eye gazing of orangutans has revealed that the amount of time spent looking at another orangutan and the area of the body being looked at seem to vary with the age of the orangutan.[25] Juveniles do most of the gazing in the group, and they look mainly at each other, with nearly all their gazes directed at each other's face. However, when juveniles look at adults, only about half the gazes are directed at the face, while the rest are focused on the limbs and body. This may suggest that certain signals are reserved for specific parts of the body. Limbs may be used for information on locomotion or hand gesturing, the face for approval or disapproval of the juvenile's actions, and the trunk of the body may be used as an area for eye-gaze avoidance (see Chapter 5). Only infants look directly at each other, and then usually at the face. They also look directly at the face of an adult. These results, presented in very abbreviated form here, suggest that communication channels may differ according to age. Adults communicate with juveniles in complex ways that include limbs, the trunk of the body and, to a significant extent, the face. Infants look at their mother's face almost exclusively just as human infants do, while juveniles observe the faces of other juveniles and may be gauging emotional state and readiness to play or fight. The fascinating aspect of these findings is that each age group of orangutans obviously sees different things and might also respond to different cues. Further work on visual perception and communication in orangutans might reveal more about their cognitive abilities.

Vocal Signals

The vocalizations of primates have received considerable attention over the years. An interesting variation from many other species is that humans and most anthropoid primates have sacrificed some of the mammalian hearing capacity in the high-pitch range in exchange for better discrimination of sounds in the middle of the hearing range. A greater ability to discriminate sounds can give rise to more complex forms of communication. But why has this occurred in primates? Researchers have argued that the gregarious nature of most primates leads to a more complex social organization and, thus, a more complex communication system. This argument has several flaws. First, there are many gregarious and group-living mammals such as ungulates (hoofed animals such as horses and antelopes) whose vocal communication might not be as complex as that of primates. Second, this explanation would discount the orangutan as a candidate for complex communication. Further, since the notion of complex communication is linked to assessments of intelligence, this would mean that orangutans cannot be regarded as highly intelligent. As we point out in Chapter 6, orangutans equal other apes in problem-solving abilities, but they are generally not gregarious.

Many studies have noted that aspects of the environment influence the vocal signals used by primates. In other words, the characteristics of their calls depend on their habitat. An open field requires one kind of vocalization and the rain forest a different kind. The forest does not readily allow long-distance vocal communication, but trees and leaves obstruct visual contact in rain forests, so individuals must rely on vocalizations to stay in touch.[26] The rain forest is an environment with high background noise, largely caused by insects and the reflection of sounds from trunks, and leaves. Foliage, temperature gradients and ground effects contribute to the absorption and distortion of sounds. Many forest-living monkeys have evolved vocalizations that are a series of rather high-pitched sounds. Orangutan males have solved the problem of communicating widely in the forest by the sheer volume of their deep, long call.

Our knowledge of the vocal communication system of orangutans is very scanty compared with the large number of studies of vocalizations in other primates. Communication in orangutans may occur purely by sound, but usually it consists of a variety of signals, including body signals such as posture, facial expression, gesturing, and touching. Orangutans may also jump up and down and emit a hissing sound to communicate their message.

The first systematic attempts to describe vocalization in orangutans were undertaken in the 1970s.[27] Very slowly, new vocalizations are being added to the original list assembled by MacKinnon. As mentioned earlier, we discovered a low-intensity purring vocalization made by Abbie when she was being held by Gisela. The context in which we heard this call led us to deduce that it conveyed contentment.[28] There may be other orangutan vocalizations that are dependent on being close or even intimate, a form of communication that we refer to as "whisper" communication. Since the personal groupings (natal groups) of orangutans rarely exceed three individuals (mother and two offspring), "whisper" communication may be a very useful form of communication for everyday life, particularly at close proximity.

Communication between mother and infant also occurs by touching, especially when the infant has been allowed to swing independently on nearby branches. The mother signals the infant to return to her body before she moves off by extending a limb, usually a leg, to touch the infant. It is conceivable that the gesture is accompanied by low-intensity vocalizations, not heard from a distance. These "whisper" calls would not be detected by most researchers observing orangutans in the wild.

We now also know that, as in many primates, vocalization in the orangutan is sexually dimorphic. Primate males usually have a different range of calls, coupled with different functions. Once the male orangutan has grown to full maturity, he is able to vocalize his presence and whereabouts by means of two different kinds of call. One is a short call that Galdikas and Insley reported first, which they termed the "fast call."[29] Then there is the so-called long call. The long call may well be the most spine-chilling and awe-inspiring roar we have ever heard in any species, except perhaps that of a lion. It is probably also the longest-lasting call of any animal. The orangutan first fills his laryngeal sac with air and then expels this air over the vocal chords, in principle not unlike a bagpipe, producing a prolonged, uninterrupted sound. When the long call is heard, life in the forest stops. The long call overrides the deafening cacophony of the insects. It echoes and hovers between the trees, filling the entire environment. It can be heard over long distances.

Opinions vary as to the meaning of the long call. Some field researchers see it as an unambiguous message by an adult male to keep other males away from his terrain. This function has been observed in other species, such as the siamang gibbons.[30] The call may serve as a challenge to deter others from getting too close.[31] During his fieldwork, MacKinnon heard over 200 long calls but witnessed only one male-male encounter,[32] so the long call may reduce aggression between males. Others argue that the long call acts as a locating signal for sexually receptive females.[33] In our experience, the long call had an impact not just on us but on many other species as well. The macaques scurried about nervously, younger orangutans shifted higher into the canopy, and we stood transfixed. Perhaps the male orangutan is the king of the forest, and his long call announces to everybody that he is about.

The "fast call" is similar to the long call but shorter in duration and appears to be used only in special circumstances. In their fieldwork, Galdikas and Insley established that only just over 6 percent of long calls receive any kind of vocal replies. Of those calls that were answered, over 60 percent were fast calls.[34] They thought that the fast call was given after conflict or some form of contact.

We have heard the fast call in a zoo. The adult male had been forced to remain indoors that day, and on several occasions, he could be heard making a deep bellowing sound, similar to a fast call. It is not clear whether he bellowed because he was dissatisfied and disgruntled at being prevented from going outdoors or whether he was attempting to communicate a message to the females out in the yard. Not all vocalizations convey a message or are intended for someone else. The question is how to distinguish whether a vocalization is meant for communication in orangutans.

Apart from the overwhelming but infrequent calls of adult males, orangutans are rather silent, unlike most forest-dwelling primates. Loud vocalizations are rare, apart from the spine-chilling calls of the male. Orangutan youngsters scream when distressed, but in general, they are quiet forest-dwellers. This is one reason they are so difficult to locate in the wild, with its rich and diverse foliage. They move noiselessly and, as has happened more than once to us and to many other researchers in the field, you can literally walk into an orangutan who seems to materialize from nowhere. Once Clementine suddenly appeared in our path, a mere arm's-length away, and we had not seen or heard her approach. We have not heard a sound from Clementine (an adult female; Plate II) in all of our many encounters with her.

Among monkeys, alarm calls are more highly differentiated than those of the great apes, including orangutans.[35] But then, orangutans and the other great apes are less vulnerable to predators and may not need an elaborate warning system. And "staying in touch," as group-living monkeys do by vocalization, might not be such a high priority for orangutans, who are contented with only occasional encounters.

So much about vocal communication in orangutans is not yet fully understood. Further study may reveal that their communication is highly complex and reliable. It is possible that orangutans have not developed vocal communication skills to the same extent as group-living primates but that, instead, they have developed visual signals for use in close encounters.

Sign Language in Orangutans

So far, we have discussed the signals used by orangutans to communicate with members of their own species. If we want to know what they really think, other ways have to be found because we cannot speak "orangutan." A ground-breaking step was taken by American researchers in the 1960s when they taught apes to use American sign language. Sign language was chosen because the vocal apparatus of apes is not constructed for speaking.[36] Recent studies with chimpanzees have used plastic tokens or computer lexigrams instead of sign language.[37]

Fewer sign-language projects have been conducted with orangutans than with chimpanzees, but some orangutans have been taught to use American sign language. Gary Shapiro of the Orangutan Foundation first conducted a fifteen-month study with four juvenile orangutans, and later with two orangutans, at Camp Leakey.[38] The most comprehensive study was carried out by Lyn White Miles of the University of Tennessee, who taught American sign language to the orangutan called Chantek.[39]

Chantek lived in close contact with his trainer and other human carers and learned 140 signs. He also invented extra signs. His vocabulary increased over the period of the study, and each day he used regularly 30–60 percent of his existing vocabulary. The rate at which he acquired new signs was marginally better than that found previously in the language-trained chimpanzees. Almost 40 percent of Chantek's communication was spontaneous, reflecting his own decision to communicate.

Very early in his training program, Chantek learned the sign for UP and used it to ask to be picked up. He was well motivated to learn this sign as he was always wanting to be picked up. In contrast, he took a long time to learn the sign for DOWN, consistent with his lack of motivation to be put down. He was also quick to learn the signs for different types of food and drink, as well as MORE and HURRY. Shapiro also found that his orangutans acquired the signs for food items with greater ease than other signs.[40] As early as the second month of training, Chantek began to combine signs into sequences, such as COME-FOOD-EAT or GIVE-BANANA.[41]

Signing may involve the use of signs that have been learned simply by associations with objects. It is much more difficult to know whether the ape understands the meaning of the signs. Chantek demonstrated the ability to do this by signaling exactly what he wanted to do. To give one example, when he wanted to go for a ride he signed CAR-RIDE and pulled his carer to the parking lot. We are reminded of BJ at Sepilok when he took hold of Lesley's hand, gestured down the path with his head, and pulled at her arm to get her to go with him. The importance of Chantek's performance is that he did not mimic signs mindlessly. He initiated their use effectively and spontaneously, and it was clear that he had something to say that he wished known.

Miles also tested his mental development. In a test involving symbolic play, language comprehension, and tool using, Chantek performed at the level of a five-year-old child.[42] This indicates that Chantek had more than just rudimentary communication skills in our human system.

Sign language has allowed us a rare insight into the mind of orangutans and other great apes, but we may still not know quite what the orangutan's real capabilities are. Sign language and laboratory experiments are artifacts of human culture and human society. Some experimenters purposely wish to make comparisons with humans and they ask "How close are the great apes to us, really?" In vocal communication, the tentative answer has to be that they are farther removed from us than most of the most sophisticated birds. Birds can learn to talk. Some birds can learn to speak in human language in a way that suggests meaning, as our galah does or Irene Pepperberg's African gray parrot Alex does.[43]

It may be that apes communicate mainly by visual and other signals rather than vocalizations. We humans have a problem with this because of our genetic closeness to apes. From an evolutionary standpoint, researchers hoped that the great apes would unlock the secret of how we moved from vocalizations to human speech, and countless training sessions have been conducted with chimpanzees to see whether the great apes hold the key to our language development. But the great apes do not have the vocal apparatus to speak. We cannot quite decide what to do with a species like the orangutan (and the other great apes), which is obviously intelligent but does not have speech. All the same, there is now plenty of evidence that orangutans learn a great deal and can solve many and difficult problems; they may be able to acquire a rather sophisticated understanding of humans and, if we are so inclined to test it, even of our language.

8

MATING, SEX, AND DIVERSITY

In orangutan society, feeding and sex are the two most obvious *social* behaviors of the species, and since orangutans are generally not very gregarious, these behaviors are often the focal points for coming together in the first place. Interestingly, human societies have also developed their most elaborate rituals around these two activities eating/drinking and sexual encounters. In some cultures, etiquette demands that food is a very private and modest occasion, not for the eyes of the public, while sex can be practiced in the presence of others. Most societies today, however, have opted for public eating and private sex. And, in general, so have the orangutans.

The Sexual Players

Orangutans are sexual beings, although they do not usually get a chance to practice sex with a partner until they are at least adolescents. But as a large body of research on human sexuality has shown, the reproductive organs become sexually responsive quite early, and this is also true of orangutans. Orangutan male infants have been observed to have their first erection around the age of thirteen months (i.e., as babies) and to begin to show active interest in their own genitalia and those of others by the time they are sixteen months old.[1] In Chapter 4, we described an eighteen-month-old male infant investigating his mother's genitalia for over half an hour. It was an exhaustive investigation, and it gave him an erection. He kept watching his penis as it so obviously acquired a life of its own and seemed quite frightened by it. On several occasions, he tried to flee after he had seen it move.

Harrisson saw young males masturbating frequently at the age of four and one-half months.[2] We have seen no reports on masturbation by infant or juvenile females, but it would be reasonable to assume that there are no marked sex differences in this behavior. Among orangutans, active sexual practice with a partner begins in adolescence for both males and females.

Curiously, the stages of life into which orangutan development has been categorized are different for males and females. Females aged five to eight years are called adolescents. Sexual maturity is usually reached in this age group, with average onset at around years seven and eight. In the case of males, however, we speak of subadults, a category that is seen to extend from eight to fifteen years. The age range of subadulthood has been described differently by different observers, ranging from eight to thirteen or fifteen years[3] or from ten to fifteen years.[4] In captive orangutans, fertile matings by subadult males have occurred as early as age six, although it has been noted that the sexual behavior of subadult males has the least reproductive success.

Sexual maturity in females is attained when the menstrual cycle is fully established. The menstrual cycle in orangutan females is almost identical to that of humans. It may vary in length from twenty-two to thirty days,[5] and this determines the number of cycles per year. The onset of menstruation varies enormously between individuals, and where the orangutan lives is a factor. Zoo environments have at times produced quite atypical patterns: For instance, the earliest onset of menstrual cycling ever described in a zoo orangutan was five years of age and the oldest eleven years of age.

After the age of eight, female orangutans are considered to be adult (and perhaps they are). Technically, they can reproduce, and often do so in zoo environments. In their natural habitat, however, orangutan females often do not begin to reproduce until the age of fifteen and males at the age of twenty.[6] There can thus be a gap of as much as seven years between sexual maturity and first pregnancy. Menopause usually occurs when the female is in her forties. There has been a case of one zoo female who menstruated until the age of forty-eight, but this was considered very late.[7]

There is a difference in secondary sexual markers between males and females, and one marker, in the male, is very obvious indeed. The male orangutan *(Pongo pygmaeus pygmaeus)* grows the very pronounced cheek pads that we have mentioned before, as well as a full laryngal sac that enables him to make calls that are audible for many kilometers. However, the onset of sexual maturity and of fully developed secondary sex characteristics is context dependent. The development of full cheek pads can be delayed merely by social conditions. It requires only the presence of an adult male in the subadult's vicinity to arrest full development of the cheek pads. In one case, cheek-pad development was suppressed until the age of nineteen.[8]

Another dramatic example was seen in a zoo environment where the adult male was removed and the subadult male left behind. The cheek pads of the younger male developed immediately.[9] This is puzzling insofar as cheek-pad growth does not coincide with sexual maturity. We know from other primates, including New World monkeys, that female sexuality and ovulation can be suppressed in the presence of a dominant female, but cheek pads are not primarily related to sexual maturity, although they may relate to reproductive success.[10] Further, the cheek-pad compartment and the laryngal sacs are not unique to males. They are present in both sexes at all ages, but they develop fully only in males.[11]

The Mischievous Adolescent Male

We have met quite a number of subadult males in both Sabah and Sarawak at the rehabilitation sites. No two male subadults are alike, so it would not be prudent to speak of them as if they were all the same, but as a group, we think of them as being akin to late-teenage boys. They seemed to have a few traits in common. They all tried to measure their strength and were, on the whole, far more dangerous than the rehabilitant females. They were irascible, feisty, and often mischievous. They loved playing tricks and setting ambushes for people. They snatched handbags, pulled on tripods, and generally enjoyed demanding objects from unsuspecting human visitors (see Chapter 3).

FIGURE 8.1
Raja, a streetwise and impressive orangutan. We stayed well out of his way.

All of them, particularly Raja, a twelve-year-old subadult male, tried to "play it cool." The ground rule seemed to be: Show as little facial expression as you possibly can and give nothing away. There may well be a very good reason for this strategy of subadult male behavior. We could imagine Raja playing the role of an aspiring gangster boss in 1920s New York. The cigar was missing, but he was very convincing in what he was doing. He looked dangerous, unmoved, and always on the verge of being bored. Yet his eyes betrayed him, shining, fast-moving, judgmental (it seemed); he was on his guard and altogether "in control" (Figure 8.1).

Subadult Male Strategies for Sexual Gratification

Subadult males are the most promiscuous of any orangutans of either sex or any age. Their sex life is well developed, and they will try any means to obtain an opportunity. But how does a subadult approach the gravely important matter of sex? Since there are no clearly identifiable family rules and no visibly present group members to consider, as in gorilla and chimpanzee society, sexual behavior in orangutan mode is individualistic, capricious, and leisurely. There are no males running around, as in some monkey troupes, to check whether a female is in heat, and there are usually no squabbles over territory or dominance, except in the most rudimentary sense, so what do they do when they are not "hanging around"?

We once saw Raja flirting with Jessica. He was some distance away and had obviously spotted her, but not once did he look at her directly, nor did she look at him. He sat there, appearing to mind his own business, but we noticed that whenever she seemed to look elsewhere, he edged a little closer to her. After some time, he was sitting right next to her. Both were on the ground, both casually holding on to a branch above them. Now he moved his hand closer until the side of his hand touched hers. He was chewing some branches and seemed to care little about the proximity of Jessica. Only the quick movements that he made when not in her visual field betrayed that he had a strategy for achieving something, and that was very focused indeed. This took almost all afternoon and eventually we had to leave, but we saw in the light of the late afternoon that he managed to plant a "kiss" on her mouth. Perhaps it was not a kiss and did not have the function that we know, but his lips touched hers and then his lips wandered all over her face. Jessica did not move, and now it was she who seemed to pretend that he was not there.

Was this flirting only, or was he leading up to sexual intercourse? In all likelihood, the latter was the case. In Chapter 4, we talked about the possible father of Jessica's baby. The accepted wisdom was that it must have been Simbo, the adult male in the region. But it need not have been him at all. It could well have been Raja or one of the other subadult males. Sometimes subadult males are successful, and a female will accept them, and sometimes this will lead to a pregnancy. It is difficult to say. Subadult males will usually choose another female for company rather than a male, even if no sex occurs.[12]

Subadult males have a few other strategies to gratify their sexual appetites, and they can be quite devious. For instance, subadult males have been known to follow a consorting pair quietly and inconspicuously, and when the adult male separates from the female and, while foraging, is briefly out of range of hearing a twig breaking or leaves rustling, the subadult male intrudes quickly and attempts to mate with the female.[13]

In nearly one-third of his sexual activity, the subadult male attempts to seek gratification through rape, although he does not always succeed. We define rape here, as in human society, as a sexual act that occurs against the will of the other party. How do we know that it is rape and not just rough sex between consenting parties? Although it is easy to overinterpret animal behavior, there seems little doubt in any of the observers' reports that what they have seen is, in fact, a rape. One reason for this conclusion is the strategies used by the male. The subadult male, at first, mingles casually in a group of females and then suddenly turns on one. Rape tends to occur not by ambushing a lone female but usually within a group.[14] Rape attacks have been observed that were almost surprise attacks, following within minutes of the male meeting the female. As in human rape, the victim is usually held down on her back.

The other indication that this behavior is accurately described as rape is the behavior of the female. The ambushed females invariably scream and often struggle fiercely to get away. Sometimes, an adult female with an offspring is raped, and the offspring, often quite young, not only joins in the screaming but may attack the subadult male or at least try to pull him away from the mother.[15] As a result, not every planned rape leads to penis insertion or to ejaculation.

The Female Club Culture

The life of adolescent females is quite different from that of adolescent males. Of all orangutans of any age group and either sex, adolescent females are the most gregarious.[16] Although most orangutans spend a good deal of time on their own, adolescent females have something resembling a regular social life. They either stay with their mother and possibly her other offspring, or they go to meet other adolescent females or subadult males. Adolescent females spend a lot of time playing and feeding together or just sitting around together, appearing to have a good time.

But they are by no means virtuous little angels. They too can have a voracious appetite for sex and may solicit to gain the attentions of a male. A female may go as far as guiding the male's hand to her genitals, pursing her lips, and generally "offering" herself in suggestive postures. If he does not respond soon enough, she may be even more direct by fondling and manipulating his penis.[17]

There is no doubt that adolescence is the time for exploration of the other sex. Most of the juveniles do not fall pregnant, but some, of course, do. Females also have preferences for one male over another. When they fancy a suitor, they are not likely to let him go. At one of the rehabilitation camps, a researcher was fancied by an adult female, and she was extremely charming to him. He thought that she was just being friendly and went along with her cuddling embraces until, one day, she decided she had had enough of the dilly-dallying and heaved him on top of her. He had to be rescued. After this incident, she sulked for weeks and was never as friendly toward him again.

Adult females may not be as gregarious as adolescents, but throughout their reproductive life, they are not often alone. As we describe in Chapter 4, once she has offspring, the adult female is rarely by herself. Although her sex life seems to diminish markedly once she has taken on the responsibility of raising an infant, there are often large gaps between such commitments. Also, copulation is not just for reproduction, as it is known to occur at any part of the menstrual cycle.[18] Active sexual interest by females is not confined to their period of ovulation, as is the case in most mammalian species.

There are two outstanding features of adult female sexual behavior. First, it seems clear from all the observations of wild or rehabilitating orangutans in their natural habitat that females, adolescent and adult alike, generally have some say in sexual matters. For example, female orangutans in the wild have been seen to move away on hearing the male's long call from a distance[19] and to avoid male contact when they have an infant to care for. Adult females are not usually raped by adult males in the wild. Unlike subadult males, fully mature adult males do not rape or do not get the opportunity to do so. If the female is not interested, she will either move away or threaten the male, whether or not he is subadult or a fully grown mature male.[20]

Here, the size difference between male and female is of some importance. In Chapter 3, we dealt with the hypothesis that the large size of the male is related to the possible need for competition with another male. Perhaps there is another reason for the size difference. If the male is persistent and strong, the female can still escape by going higher into the canopy to reach branches that will support her weight of about forty kilograms but not his weight of eighty kilograms or more. The higher parts of the canopy are her realm, and there is nothing he can do about it. We wonder whether the female's habit of giving birth in a nest high in the canopy and under cover of darkness is entirely related to problems with predators (see Chapter 4). Is it conceivable that giving birth in the upper canopy at night protects her from curious or desirous male orangutans?

The second unusual feature of orangutan sexuality is that the female genitalia do not swell at the time of receptivity. Gorilla and chimpanzee males walk around inspecting their females from behind, and they will know, despite the fact that she may try to indicate otherwise, when a female is receptive. Orangutan males have no such easy task. As in humans, there are no outward and visible signs to help males judge the receptivity of females. So it appears that as in humans, making sexual contact among orangutans may well be a guessing game, based on negotiation and persuasion, unless olfaction also plays a role that we have not as yet discovered.

Orangutan Males: Hermits or Merely Antisocial?

Mature adult males show a marked aversion to each other. Once they are fully matured with cheek pouches and air sacs, they live very solitary lives, spending over 90 percent of their time on their own. When they do contact other orangutans, it is not for lasting or necessarily friendly purposes. Rarely is an adult seen together with a subadult male, and the few encounters that have been observed have been brief and limited in interaction. Other adult males are avoided at all cost. An orangutan male has calls that ensure that everyone in a neighborhood of several miles knows his whereabouts (see Chapter 7). It has never been entirely clear whether these calls ward off other males or attract females, or both. If two adult males do meet, they are not friendly. There are countless reports from the nineteenth century onward to verify that when these rare encounters occur, they are generally hostile, ranging from threat gestures to confrontation; brief but fierce fights may even occur, causing severe injuries at times.[21]

Any other form of contact with other orangutans is almost exclusively with a consorting female. Such a consort relationship may last for a few weeks at a time. Biruté Galdikas reports that nearly all of the 10 percent of social contact occurring in males is the result of sexual partnerships with females.[22] It seems, then, that the adult male orangutan should be described as an antisocial being rather than a hermit. He does seek out females and can be quite charming and gentle.

The Sexual Act

The sexual act, and the lead-up to it, finds no equivalent in the sexual behavior of other primates, except in bonobos (pygmy chimpanzees) and in humans. There are several unusual aspects of the sexual act among orangutans (and bonobos) as compared with the common chimpanzee and the gorilla. Male gorillas and male chimpanzees mount their females from behind. Orangutans (and bonobos sometimes) mate face-to-face. There is usually no foreplay involved with the other great apes (gorilla and chimpanzees), although sexual intercourse is initiated by females in almost half of all matings observed in chimpanzees.[23] In these species, mating is a brief affair, lasting only a few seconds and barely interrupts the male's vigilance in keeping order and control in his troupe.

Orangutan sexual intercourse is very different. It is an all-absorbing activity. First, there is extensive foreplay, which is unhurried and measured. Advances are made that may involve sitting next to each other and casually touching the body of the other. Subadult males have been observed to put their arms around the female[24] and to touch the face of the other and the region of the ear with their lips. Unless she actively solicits, which does happen, her passive behavior shows at least a lack of protest. Part of the sexual foreplay includes the male licking the female genitalia and fondling the anogenital region.[25]

Among nonhuman primates, the orangutan is the only one to practice sex exclusively in the frontal position. In bonobos, up to one-fourth of all matings may be face-to-face, with the qualification that more of this behavior occurs in captivity than in the wild and that a large percentage of face-to-face sexual encounters are female-female.[26] Intercourse between consorting orangutan pairs occurs face-to-face and may be executed in a number of positions. They may hang from branches, facing each other and embracing each other with the legs. In the "missionary position," the female lies on her back and the male is seated between her legs or on top of her, although not touching her body. Alternatively, the male may lie on his back against branches of a tree, being straddled by the female, who performs the thrusting motions.[27]

The time taken by orangutans for copulation is the longest among nonhuman primates, equal in length only to that of humans. The average copulation time in the Tanjung Puting area was observed to be about ten minutes, ranging from three minutes to half an hour in the wild and even longer in captivity.[28] Do orangutans experience orgasm? The answer may well be that they do. Females have been known to scream toward or at the end of the copulatory bout, suggesting that she has at least experienced an orgasmic sensation. We know that these particular screams do not occur at any other time and are hence strongly associated with intercourse.[29]

Long acts of intercourse are not without risk in the natural habitat. Danger or competition could be close by at a time when the attention is otherwise engaged. One of the reasons intercourse among most species is very brief is to keep the time of being vulnerable to predators to an absolute minimum. There is also a second important reason. Unfinished business may give a competitor the chance to step in and take over. This has been observed in orangutan society—a small adult male or a subadult male may well be interrupted and supplanted by a large adult male.[30] Such a change of partner during intercourse would hardly be possible if the mating bout lasted only a few seconds.

Sexual activity between a consorting couple may continue on a daily basis for several weeks. Thereafter, both go their separate ways. The fertile period is five to six days per cycle in orangutan females, two to three in gorillas, and ten to fourteen in chimpanzees.[31] Following the consorting period, pregnancy may be detected two to three weeks after conception. The outer labia begin to swell, and this labial swelling becomes very pronounced.[32] This is one time when sniffing by subadult males seems to occur: On sniffing the region of the swelling, they will let go of the female and not make any sexual advances. The pregnant female then goes off on her own.

Males do not take any part in parenting, nor have they been seen to function as protectors of females in case of an attack or other dangers. Female orangutans derive no benefit from their males other than sexual contact and the offspring itself.

Sex and Reproduction in Zoos

Why do we want to know exactly how and when orangutans have sexual intercourse? As we said earlier, orangutan numbers in the wild are declining, and they may be approaching extinction. This has created a role for zoos.

Many zoos today have taken on the task of trying to breed endangered species with a view to maintaining genetic diversity and preserving them as a viable species (see also Chapter 9). In pursuing this goal, reproductive biology has become an important area of investigation in the face of increasing threats to the survival of many species in the wild. The orangutan is one of the species that zoos try hard to breed, but many zoos willing to embark on such a program find themselves unprepared for the task. They need to know more about orangutans, particularly their social organization.

Until the early 1990s, very little was known about the reproductive biology of orangutans, especially that of the male.[33] That situation is changing rapidly, and the genetic diversity of captive orangutans has become an issue of some concern. Some decades ago, captive breeding stock was regularly replenished by wild orangutans. These days, accredited institutions are ethically and legally bound by international legislation on endangered species not to capture wild orangutans in order to replenish their existing genetic pool.

One specific problem of breeding in zoos is that the two subspecies of orangutans, the Sumatran and the Bornean, have often been crossbred. Now the American Association of Zoos and Aquariums (AAZA) has developed an Orangutan Species Survival Plan that recommends that the hybrids of the two subspecies should no longer be used for breeding purposes and should be maintained separately. In breeding terms, this has created a sizable "surplus" in zoo orangutans that are now of no use for maintaining genetic diversity.

Further, the "founder" (i.e., wild-caught) orangutan males of the 1960s (and perhaps later) are reaching the end of their reproductive life, and some sources claim that they have been underutilized. Many zoos have too few offspring or only hybrid offspring. The fear has been expressed that the irreplaceable genetic information of the founders could be lost forever unless action is taken now.[34]

A further problem facing breeding zoos is the provision of adequate conditions for keeping and rearing orangutans. Many zoos are not set up for breeding or lack a suitable environment to meet the essential needs of orangutans for social and psychological well-being.

As the Great Ape Project has shown in some detail, lack of attention to the behavior of great apes and to their needs has caused individual apes great suffering, early death, and, of course, little or no reproductive success (more details in Chapter 10). Orangutans are often housed together with little regard for their social or psychological needs (sex, age, relationship, and personal likings are important factors).

Animals have been bought and sold as property, not as thinking (and feeling) beings with preferences, likes, and dislikes. Zoos have often handled the breeding in the same manner. Just as we would not expect that bringing two people together in the same room would automatically result in their mating, so we should not expect this of orangutans. Yet there has long been an assumption that sexual behavior in apes is what humans imagine to be "animal-like" (i.e., instinctual, indiscriminate, thoughtless, and rough) and therefore something they will do regardless of circumstances. Nothing could be further from the truth. It is now known that even in their natural habitat, orangutans will not reproduce if conditions are not right. It has also recently been found that orangutans (and other great apes) raised by humans lack the social skills to mate.[35]

We now know that orangutan sexual behavior is complex and needs to be learned. There are rules and rituals that must be known, and there are ways of creating better conditions for breeding to occur. Experiments have been conducted with captive orangutans to find out whether choice by females and decisionmaking about time of mating are important. R. D. Nadler of the Yerkes Regional Primate Research Center at Emory University, Atlanta, observed mating behavior under two different conditions. In one testing environment, the adult male was free to enter the enclosure of the female whenever he wanted. The result was that copulatory bouts increased well beyond those known in the wild, and the contacts also resulted in rape. In the second testing condition, access by the male was restricted, and it was the adult female who determined access via a door mechanism. The number of copulations decreased substantially.[36]

The Species Protection Plan devised for orangutans and a number of other species may lead us to believe that great care will be taken in breeding them in captivity. But the new emphasis on breeding has often brought hardship to individual apes. A technocratic professional elite has taken the planning in hand to save the species in any manner it can. The underlying assumption is relatively simple: We need diversity and we need strategies to make this happen.

A human interventionist approach that has been favored in the 1990s by several institutions is to bypass natural breeding conditions. The aim is to collect eggs and sperm and make reproduction happen in a test tube. There are now several techniques to gain access to orangutan sperm (called semen "recovery"), requiring either invasive or noninvasive human intervention. There is the possibility of direct extraction under anesthesia. Another method is the rectal probe electro-ejaculation technique, which also requires anesthesia of the animal and often results in semen with a low sperm count and thus not really usable. There is electro-ejaculation by direct penile stimulation, a method that has been used in macaques, and the more direct method of "digital manipulation" and use of an artificial vagina that has been trialed in a zoo.[37] The reproductive status of female orangutans may be examined by collecting urine and measuring estrone (a noninvasive procedure). In addition, follicle stimulation and ovum collection is now possible in orangutans; these are invasive procedures.[38]

Individual apes are identified by ISBN (International Studbook Numbers). There are target founder contribution (TFC) analyses that identify specific characteristics, and attempts are made to maintain particular sets of genes across generations.[39] All matters of life and birth are discussed in terms of technicalities and reproduction, as if the species were part of a large new business venture.

In this context, concern for the individual animal is easily lost and may be dismissed as sentimentalism. The strategies may appear logical in terms of strategic planning—whether they will do anything worthwhile for the survival of the species has yet to be proven. As orangutan sexuality is a matter of great interest and reproduction of orangutans a matter of urgency, new pressures have led to new abuse.

There are orangutan females who now reproduce below the normal age of reproduction in the wild, risking a reduction of their life span and poor mothering.[40] Female orangutans, who would normally bear no more than two to four offspring, are being used as birth machines, producing up to eleven offspring. There are female orangutans who have repeated miscarriages but are forced to reproduce until they have a viable offspring. Adult males and females are housed together permanently, forcing the female to contend with a far higher rate of intercourse than she would ever choose to experience in the wild. Confined space, lack of diversion, and forced proximity, as well as the fact that orangutans in zoos are often stressed because they are being watched, lead to an increase in rape, an unusual occurrence in fully grown adult males living in a free-ranging situation.[41]

Of course, there are also examples of zoos with excellent animal husbandry of orangutans.[42] A close look at their methods reveals that the animals are treated as individuals and attempts are made to cater for their needs.

Another strategy to maintain species diversity is to move the population of great apes around from one zoo to the next or from one enclosure to another, with little thought for the impact on individual apes of changing their partner, friends, environment, or reproductive intentions. David Cantor tells of bungled and heartless dealings with gorillas Katie and Timmy.[43] Timmy, a silverback male lowland gorilla, had lived in one zoo since 1966 and had been isolated from other gorillas for thirty years. When he was introduced to females, he did not like them and did not mate with them. In 1990, a new female was put into his cage, and their liking for each other was instant. Katie and Timmy were very affectionate to each other and slept together in each other's arms. This could have been a happy ending for Timmy's lonely and unhappy life, but it was not to be, because Katie could not conceive. The decision was made to separate them and reorganize the groupings. Timmy went to another zoo, and Katie was given a new mate. Katie and her new mate fought. The new male was abusive, and she was bruised and injured by his attacks, leading to the amputation of a toe that he had bitten. Timmy, in his new environment, never produced an offspring. Most of the day, he sat on a rock and stared blankly into the air.

The decision to move Timmy seemed rational in terms of breeding requirements and the plan to "protect" the species. Under the word "protection," however, can hide immense cruelty. The decisions regarding Timmy were not humane, nor were they even vaguely related to knowledge of gorilla behavior. Indeed, the zoo director at the time said in response to the anger of the public:

> It sickens me when people start to put human emotions in animals. And it demeans the animal. We can't think of them as some kind of magnificent human beings; they are animals. When people start saying animals have emotions, they cross the bridge of reality.[44]

There is no need to treat apes as human beings, but there is a great need to understand their behavior better so that decisions are informed by knowledge of the species rather than by prejudice and ignorance. Sexual behavior is one very important variable in the viability of the species, both in the wild and in captivity. Some zoos are now realizing that apes cannot be moved about at will without causing distress (which lowers reproductive success) and without creating individuals who are so psychologically disturbed that their role in parenting becomes useless.

What is the point of achieving a diverse gene pool while at the same time reducing those orangutans, who are still living, to a state of dysfunctionality, both physically and psychologically? There is no replacement for psychological well-being. Even sexual behavior does not come "naturally."

Part III

THE FUTURE OF THE SPECIES

9

IN HUMAN COMPANY, CAPTIVE OR FREE

Orangutans might never have chosen to be in human company, but that is where many of them are. The reasons for such contact often went seriously against the interests of the species. Quietly, the human predator has stalked the orangutan almost out of existence. If there is to be a halt to their elimination, it is now in the hands of humans. To save them requires a specific attitude and urgent action.

Becoming a Political Animal

In 1988, Booth entitled a paper on the reintroduction of certain species into the wild "Reintroducing a Political Animal."[1] The term "political animal" is wonderfully appropriate for the orangutan today.

The most important criterion of the "political" animal is, of course, that it must be endangered. There is a price on its head, and specifically, it is not just on a subpopulation or group but on the species as a whole. This fact was actually understood very early. The only two habitat countries of the orangutan, Indonesia and Malaysia, both instituted some form of protection for orangutans many years ago. Indonesia forbade the killing of orangutans in 1925 and then their export and trade in 1932 and 1933. In 1937, orangutans were declared endangered in Indonesia. In British North Borneo (now called East Malaysia, with its two provinces of Sarawak and Sabah, since Malaysian independence in 1967),[2] it required time and effort before similar actions were taken. Largely through the remarkable dedication of one woman, Barbara Harrisson, who then lived in Sarawak, the plight of the orangutan was made widely known. She achieved the enactment of legislation to prevent the killing, keeping, or capturing of orangutans. Capturing them was declared illegal in Eastern Malaysia in 1963. At first this legislation existed only in Sabah. Harrisson also managed to establish the first rehabilitation station for orangutans, and this opened in Sarawak in 1961 (see Table 9.1).

But unfortunately, at that time the rest of the world did not know or care about the red ape in Asia. The orangutan had suffered an "image problem" for a long time, so it was not an immediate candidate for becoming a "political animal." Capture, poaching, sales, killings, and trade continued almost as if nothing had happened to protect them.

Bad Press

The bad press about orangutans started with reports by adventurers and naturalists. In 1838, for instance, one writer described orangutans as slovenly and useless creatures:

> Their deportment is grave and melancholy, their disposition apathetic, their motions slow and heavy, and their habits so sluggish and lazy, that it is only the cravings of appetite, or the approach of imminent danger, that can rouse them from their habitual lethargy, or force them to active exertion.[3]

This attitude persisted for over 100 years. It was echoed in 1924 in a scientific paper by Sonntag, who proclaimed: "The Orang is the least interesting of the Apes. It lacks the grace and agility of the Gibbon, the intelligence of the Chimpanzee and the brutality of the Gorilla."[4] And even in the 1960s, Reynolds claimed of the orangutans that there was "nothing very spectacular about them."[5] The orangutan was thought to be less suitable for experiments,[6] less capable of problem solving,[7] and less skilled in manipulating capacity and ability than other apes, especially the chimpanzee.

TABLE 9.1 Establishment of Orangutan Rehabilitation Centers

Founding dates	Location
1961, 1977,	Bako National Park, replaced by Semenggoh in Sarawak, East Malaysia
1964,	Sepilok, near Sandakan, Sabah, East Malaysia
1971,	Ketambe, Gunung Leuser Reserve, North Sumatra, Indonesia
1971,	Tanjung Putting, Kalimantan Tengah, Borneo, Indonesia
1973,	Bohorok, Lagkat, Gunung Leuser Reserve, North Sumatra, Indonesia
1990,	Tanjung Harapan, Tanjung Putting Boundary, Kalimantan Tengah, Indonesia
1991,	Semboja, Wanariset Research Centre, Kalimantan Timur (Borneo), Indonesia

A variation was played out in stories and drawings that told of orangutan males who abducted fragile English ladies and raped them. To complete the negative imagery, drawings of ugly and frightening orangutan faces circulated in the polite society of Europe and no doubt generated dislike and fear. The orangutan exemplified what Thomas Hobbes had described as the ultimate outcast image of "man"—devoid of ordered social existence, living a life that was short, nasty, and brutal.[8]

A second way of ignoring the plight of orangutans stemmed from the pragmatism of a new class of entrepreneurs who went to the colonies to make money—if not by procuring natural resources, then in plantations. In Borneo, monocultures were introduced, and these began to dislodge the native fauna. A diary kept by one of the colonial wives affords a glimpse of the attitudes of the day (1894):

> Diary entry 15 August 1894:
>
> Later Mr. H. went, and after 2 1/2 hours absence arrived saying the Sooloos had got the poor beast but it was not dead so they brought it bound hand and foot and placed it before the house. I felt so sorry for it; for it had been shot and had also had a fall from a tree, but it was very brave and patient not making a sound altho' it must have been suffering severely.
>
> W. said it was a female of the large species. As it was tremendously powerful and too large to keep in captivity and was wounded and also

> they thought had an arm broken it was the most merciful thing to kill it, but it would not die, they had to shoot it three times before they succeeded in dispatching it. A horrid affair but orang utans can't be allowed to roam about at will devouring the sugar cane.[9]

The life of orangutans near human habitation or plantations has not improved much to this day. They are shot if caught anywhere near a plantation.

Positive Marketing

It was not until the 1960s and 1970s that more positive and sober voices were heard. People began to learn of Barbara Harrisson's work, and also that of the American primatologist Duane Rumbaugh. Jürgen Lethmate, the German primatologist, carried out behavioral studies between 1974 and 1977 that proved conclusively that the chimpanzee's intellectual superiority over the orangutan was a myth.

But for the orangutan to become a "political" animal, much more was required than scientific work by specialists. It needed the media to step in to help the orangutan gain world attention. The media began to oblige in the 1990s and marketed the orangutan as an attractive and "gorgeous" animal. Articles such as "Born to Be Mild,"[10] describing the orangutan as "trusting" and "inquisitive," "endearing, loyal pets," and "one of the world's most delightful and rare animals," effectively presented another interpretation of the orangutan. It was almost immaterial whether this view was any more correct than the earlier negativity. The 1990s also saw a spate of films on orangutans, marketed by leading television documentary makers. Probably the most talked about was the sympathetic orangutan documentary presented by actress Julia Roberts in 1997.

Then there were newspaper reports of the cruel transportation of orangutan babies that made international headlines and roused the sympathy of many. After orangutan trade had been declared illegal, transport conditions deteriorated because the live cargo had to be hidden, often exposing the orangutans to inhumane conditions. It is said that for every five orangutan babies, only one arrives alive at its destination. In the 1990s, smuggling of orangutans continued, even to zoos such as Belgrade and Moscow,[11] well after such trade had been declared illegal by the Convention on International Trade in Endangered Species of Wild Fauna and Flora (CITES). Illegal trade, in fact, boomed throughout the early 1990s. In 1992, in Taiwan, orangutans were being traded for the equivalent of $6,000–15,000 each. Between 1988 and 1990, an estimated 1,000–3,000 orangutans were thought to have been smuggled to Taiwan alone.[12] Observers of illegal trade fear that the illegal routes will shift to other countries that are not part of the CITES treaty, particularly as the lucrative live-animal trade is now said to be linked to the drug trade.[13]

The most famous case of illegal transport of orangutans was the "Bangkok Six" in 1990. Six infant orangutans made a trip from Borneo to Singapore and then to Bangkok, stuffed in poorly ventilated wooden crates without food and water. One crate was accidentally placed upside down, and for the duration of the trip, the babies traveled head down. They were discovered at Bangkok airport and eventually handed over to the Wildlife Fund in Thailand. The media covered their story. Photos of the infants showed fragile and bewildered babies. The orangutan was on the front pages of the world media for the first time, finally on its way to becoming a fully fledged political animal. Yet it was not quite there, because the world never learned that despite every effort to save them, all except two of the "Bangkok Six" died from the effects of the journey.[14]

Why did the media take up the plight of the "Bangkok Six"? Clearly, by the 1990s, orangutans made a "good" story, precisely because the species was endangered and the infants were photogenic. But there had also been controversy, politics, and marketing from several different directions. One was tourism. In 1990, Malaysia started a campaign called "Visit Malaysia Year" and adopted the orangutan as its mascot. In 1991, more than 7 million tourists visited Malaysia, and the country has maintained a rising tourist profile ever since.[15] The orangutan was placed at the head of this tourist campaign.

Tourism in Malaysia covers a wide range of activities, as elsewhere, but hardly any other country in the world has developed ecotourism on a scale as large as Malaysia. Information on Malaysia's national parks and preserves is now available on the Internet through the Malaysia Tourism Promotion Board, detailing important information about the parks, how to get there, and what tourist facilities are on offer. In Sabah, the Kinabalu National Park, Niah National Park, the Tunku Abdul Rahman National Park, the Gunung Mulu National Park (caves), and the Sepilok Rehabilitation Centre are popular tourist spots. In Sarawak, Semmengoh Rehabilitation Centre and Bako National Park feature widely as tourist attractions.[16] The existence of some national parks and wildlife reserves is now secured by the income they generate.[17]

One of the main attractions of ecotourism is the orangutan. Malaysia has chosen it as its mascot, although the orangutan is not unique to Malaysia but occurs also in the Indonesian part of Borneo and in Sumatra. Travel brochures advertising Borneo invariably carry photographs of three things: local people, usually in full ceremonial outfit; a native animal, in this case orangutans, usually baby ones; and one or two scenes of undisturbed virgin rivers and forest. They are meant to convey the unusual (orangutan), the indigenous (people), and the desirable (wilderness) and to bring into reach the inaccessible.

This strategy of marketing the uniqueness of Malaysia via the orangutan has worked. It seems to have become "chic" in the Western world to have seen orangutans or to have become acquainted with them in person. There is scarcely a travel catalog or newspaper in Europe, Australia, or the United States that does not at some time feature a photograph of orangutans or comment on them. The charm of orangutans has even trickled down to news and gossip columns in rural Australia,[18] and Sepilok is the most frequently mentioned site for a special experience featuring the cute humanlike images of baby orangutans. These days, orangutans are referred to as "gentle giants,"[19] and comments confirm, in typical anthropomorphic style, that they are not all that different from us, making it "hard to believe we were watching baby orangutans and not humans."[20]

Belated as this interest in orangutans is, it cannot be divorced from the Malaysian marketing drive to boost tourism.[21] The world now knows about the decline of the orangutans, and the world responds largely by taking up the offer to catch a last glimpse of this tropical paradise before it disappears. Discovering "paradise" at the very time it is so nearly lost belongs to the contradictions of development and symbolizes the environmental tragedies of the twentieth century.

Another step forward in the 1970s was the increasing strength of the environmental movement. By the 1980s, it had gained sufficient force to have its concerns broadcast on commercial television stations. The environmental movement was not directly involved with orangutans, but it had serious concerns about the rain forests of the world.[22] Rain forests in the Congo, in Borneo, and the Amazon were identified as the oldest, most valuable, and most diverse in the world. Attention began to focus on these forests and on attempts to save them.

Slowly, the importance of saving the rain forests filtered through to the general public, and it became widely understood that their specific ecosystem functions as a kind of "air-conditioning" and filtering system for the whole world. Moreover, they contain the greatest concentration and diversity of flora and fauna on earth. Disappearance of the forests, it was realized, could change climatic conditions, not just locally but worldwide. By this time, however, it had also become fashionable to speak of sustainable forestry and selective logging, so public concern for rain forests acquiesced for a while. Along with it, interest in orangutans declined.

What Help Is There for the Orangutan?

Indonesia and Malaysia both put in place legislation to protect the orangutan but, ironically, not its habitat.[23] On the contrary, both countries increasingly used their forests for commercial logging. In 1986, the Indonesian government decided to open up Kalimantan, largely by way of timber concessions and plans for plantations. MacKinnon, foreseeing disaster, calculated that only about 2 percent of the original orangutan habitat was protected.[24] Orangutans could not be sustained as a species in such reduced areas of rain forest. Even without further reduction, the risk of genetic erosion in orangutans was already very real. A working party found that in the existing reserves, the population of orangutans was well below the "genetically safe" population size.[25] Because of the fragmentation of their populations, nowhere do orangutans today live in breeding groups of the estimated "genetically safe" size.

Added to this, orangutans do not take kindly to overcrowding, even though they may appear to be adapting to new conditions. If too many orangutans are relocated to one area through logging or agricultural developments, they slow down reproduction or even stop. They respond by self-culling, thereby hastening their own demise.

By the 1990s, there was enough factual material on the international table for even the most skeptical minds to see that the situation was grave. For instance, the World Resource Institute established that rain forests worldwide were being decimated at the rate of 100,000 square kilometers per year. The damage that even selective logging does to rain forests was also more widely publicized. In 1993, the Working Group on Orang Utans, a specialist group consisting of members of the Indonesian Forest Protection and Nature Conservation and the Captive Breeding Specialist Group and Species Survival Commission of the International Union for the Conservation of Nature and Natural Resources, compiled a report called *Orangutan: Population and Habitat Viability Analysis Report* and made the following recommendations:

1. The major threat to the orangutan population level is the loss of adult females and the low rate of population increase which is natural for this species.
2. Given the life history of orangutans, continuous vigilance and strengthened enforcement of existing laws is required to protect existing populations because they are unable to withstand significant levels of removal through poaching.
3. The major threat on orangutans is habitat loss as well as degradation and fragmentation, especially in lowland forest.[26]

In Sabah and Sarawak, the Bornean rain forest is almost a thing of the past, and everyone knows that where there is no habitat, there is no place for the orangutan. The circle has been closed. The orangutan has finally, it seems, become a fully fledged "political animal." But can it be saved, and can the rain forest be saved?

Zoos

Some believe that orangutans can be saved through captive breeding programs. If well managed, captive breeding programs can help to maintain the genetic diversity of a species and prop up their numbers. Jeffrey Black argues that captive propagation essentially buys time in which management errors can be corrected, habitat restored, and, if possible, overexploitation minimized.[27] Such breeding programs are almost exclusively undertaken by institutions, such as zoos. Reliance on zoos for breeding programs has increased greatly.

In recent decades, zoos have begun to play an important part in the overall fate of species, many turning their attention to conservation issues. Some even perceive their new role to be that of a "Noah's Ark"—a refuge for species that are endangered or extinct in the natural environment. This new self-understanding of the role of zoos has been criticized as much as it has been praised.[28] The question is whether zoos can ever really hope to fulfill this role of a rescue mission for endangered species.

With few exceptions, orangutans have not adapted well to living in zoos. The International Studbook records reveal that between 1946 and 1994, a total of 1,975 orangutans had been held in 810 zoos. In the period 1959–1965, the average life span of orangutans in zoos was ten years.[29] Part of the problem (as for other animals in zoos) is transportation. Zoos exchange animals, and animals for breeding are often sent halfway around the world, to different climate zones, different seasons, different vegetation. Of the 1,278 orangutans that were transported from one zoo to another between 1946 and 1994, more than one-third never made it to their destination. They died in transit. A further 16 percent of these orangutans died within a year of arrival.[30]

The survival rate for orangutans in zoos has never been good, but it was particularly low before the 1970s, due in part to poor housing and lack of any stimulation. Up to the 1970s, when orangutans arrived in zoos as adults, they could look forward to staying alive for about another ten years, not taking into account their age on arrival. When transport problems are removed from the statistics, the average zoo life span for an adult orangutan (including every orangutan over the age of five) during the entire postwar period (1944 to 1994) was eighteen years. This means that, on average, most orangutans died young and in their prime. A healthy orangutan has a life expectancy of forty to fifty years. The world's oldest zoo orangutan, named Mawas, from Perth Zoo, died in 1997 at the age of fifty-six.

Psychological problems (often demonstrated by stereotypical behaviors) have been noted in 11 percent of all captive orangutans. Every fifth orangutan female dies after the birth of an infant, and nearly 40 percent of orangutan females reject their infants.[31] These are not good figures, and they detract from the belief that zoos are suitable places for a rescue mission.

Moreover, zoos themselves now face a crisis in preserving the genetic diversity of orangutans since they can no longer rely on replenishment stock from the wild. The "farming" of animals means they have to rely on reproducing from captive populations, often a psychologically damaged group. Most zoos in the Western world exhibit species from countries in the developing world, so the species are not in their habitat countries and the degree of "artificiality" is enhanced.

The idea of a "Noah's Ark" is flawed in another way. Captive breeding programs are costly and labor-intensive. Not every zoo can take all endangered animals, and all the zoos put together could not make special provision for all currently endangered species. In terms of cost, benefit, space, and resources, this is currently not manageable. Thus, zoos have to choose the species they think ought to be saved and can be accommodated. The process of selection becomes controversial. In the political jargon of breeding programs, we now speak of "flagship" species (i.e., those that will gain public approval and therefore the financial support that is required for their maintenance). Zoos as "Noah's Arks" are thus, of necessity, politically motivated and selective. This is of concern even if we could be reassured that orangutans would be among the selected species.

There is no known correspondence between what is considered to be good performance of animal husbandry in a zoo and good performance of a species in the wild. The problems of zoo environments include the risk of loss of fitness due to lack of competition, loss of skills and knowledge, loss of behavioral repertoire, and loss of ability related to recognition of food, predators, and plants in the natural environment.[32] Captive breeding programs face problems similar to those of the rehabilitation programs for wild-born orphaned individuals. Human-raised orangutans often reject their own species (and offspring). Cody was an orangutan hand-raised by humans. When he was introduced to another orangutan:

> Cody . . . was stricken with terror on first beholding another orangutan. The hair stood up all over his body. He recoiled in fear and hid behind his

human "parent," clinging so hard that he left marks. The placid orangutan who frightened him so much happened to be his own mother.[33]

Another psychological problem is boredom. There are no cages large enough for orangutans, who have a natural-habitat requirement of several kilometers. Indeed, there are no cages large enough for any large mammal. Being enclosed in this way may result in stereotypical behavior (highly stylized behaviors such as rocking, moving up and down the cage, odd repertoires of distorted limb movements), self-mutilation, or, at the very least, utter boredom, sadness, and despondency. The escape of Indah, an adult female at San Diego Zoo in 1993, exemplified this well. When she managed to get out of her enclosure, she did not attack people (as some expected) and she did not flee (as others were afraid she might), but rather went to a waste bin on the viewing deck and started dismantling its contents. She examined each item, indulging her curiosity for once.[34] Masson and McCarthy make the point that Indah must have been thinking about the contents of the bin as she peered at it daily from her cage. The modest pleasure of doing something rather than nothing highlights the tragedy of their lives.

Orangutans in the wild have a myriad of new impressions and new situations to deal with on a daily basis. Their minds are kept active. For Indah, the waste bin was a brief reprieve from the drudgery of captivity and the utter lack of stimulation. Even so-called enriched environments, while much better than bare cages, are limited. An enriched environment may contain toys or varied locations for hidden food (e.g., syrup in a tree hollow) but after a few days, the novelty wears off and, usually, the same boredom sets in unless there is a constant change of toys and structures. Zoos rarely have the resources or staff to provide what captive orangutans need daily.

These examples show that the breeding outcomes in a zoo are likely to be a very inferior (zoo) version of the wild state. Captive breeding programs inadvertently promote the end of natural selection, replacing it with a selection for captive environments.

Health risks are also exacerbated in zoo environments. Many animals in captivity, especially orangutans, suffer from chronic stress. Stress is known to result in high levels of pituitary-adrenal activity, which can have inhibitory effects on reproduction as well as reduced immune responses, stunting of growth, and problems with digestion.[35]

The story of Noah's Ark has an ending that we cannot expect zoos to match. After all, at the end of the storms and the floods, once the waters had subsided, Noah was able to open the door and let all the animals go to multiply and live happily ever after. Open the door for captive-reared orangutans and release them into the natural environment, and their future is sadly predictable. They will not recognize the forest or know how to cope with it. They will be total strangers within it. In other words, animals kept for any length of time in captivity have every chance of becoming extremely poor candidates for release. Most of them are destined to remain in captivity. This is the case for orangutans in zoos.

Rehabilitation

Many people now feel strongly that the orangutan deserves to live in its natural environment and not be held captive, for any reason. How admirable such sentiments are and how very difficult to implement in current environmental conditions!

In Malaysia and Indonesia, a number of rehabilitation centers now exist to service a troubled wild orangutan population. Most of them are on the island of Borneo, but there are two in Sumatra (see Table 9.1). The purpose of these rehabilitation centers has not always been clear. They provide medical help to orangutans, accepting particularly infants and juveniles whose mothers have been killed, and confiscate orangutans that are still being kept illegally in private homes.

Part of their brief is to rehabilitate and then release orangutans that have been declared fit (in social skills as well as in health) to go back to the wild. This outcome has often been achieved in cases when adult orangutans have been treated for fractures or other short-term illnesses. Once the fracture has healed, the orangutan can go back to its natural habitat. This temporary support is where rehabilitation centers play a valuable role. Provided all veterinary procedures are followed, little damage is done.

The other valuable role that rehabilitation centers have taken on is training wild-born infant and juvenile orangutans for release back into the wild, but whether they are actually succeeding in doing that has been debated for years. Some writers are critical of long-term rehabilitation attempts, claiming they are a dismal failure because they involve too much guesswork, mainly because so much of the orangutan's biology, psychology, and social behavior is unknown. It has been said that rehabilitation centers are mismanaged in terms of orangutan behavior and that the animals have not benefited from their experience.[36]

These criticisms are not surprising, for two reasons. When rehabilitation centers were first opened, mostly in the 1960s and 1970s, the orangutan was virtually an unknown and unstudied species.[37] Moreover, the study of animal behavior in the context of release programs was usually considered irrelevant. This has to do with the perception of many people, professionals included, that animals somehow know most things by instinct[38] and that, therefore, it was only important to consider ecology and health as preconditions for release. Behavior (i.e., their psychology and need for learning) was often only of marginal interest to rehabilitators. Secondly, staff in rehabilitation centers tend to come from all walks of life, and few are trained in biology or ethology (animal behavior). There may always have been goodwill, but it was often dilettantish and possibly did more harm than good.

The goal of rehabilitation centers should be release. But that goal can be unachievable at times and, even in the best cases, is increasingly more difficult to implement because rehabilitating orangutans, especially those that have been abused in illegal captivity, are very damaged individuals. Some orangutans coming into care are in such poor psychological condition that their release back into the wild is out of the question. Many are so young that they have to be hand-raised by humans. All data from captive populations over the last fifty years have shown clearly that hand-raised orangutans have a substantially lower survival rate than mother-reared orangutans even in captivity. The problem is magnified many times over in attempts to reintroduce a hand-raised orangutan into the natural environment. Hand-raised orangutans also usually fail as parents,[39] so that the final test for successful rehabilitation (i.e., successful reproduction) will hardly be met.

Of lesser importance to rehabilitation problems now are the health issues arising from contacts with humans, although these do have implications for the release of orangutans. Orangutans can contract just about any disease that humans can suffer. Humans are disease carriers and pass on illnesses that can strike down an orangutan whose immune system may not be adapted to deal with the diseases as successfully as humans. Influenza, measles, and malaria are just some of the diseases lethal to orangutans. One of the chief killers, until recently, has been tuberculosis, which occurs relatively rarely in the wild but is quite common in captivity. It is caught through frequent exposure to humans.[40] For instance, there was a planned release of orangutans in Indonesia that had to be canceled when it was discovered that some had been infected with tuberculosis from humans.[41] Rehabilitation centers now usually have this problem under control through regular inoculation and testing.[42]

Another serious threat is the human herpes virus. Taiwanese nightclubs quite often "employ" orangutans as attractions. Some were eventually rescued and taken back to Malaysia but were found to have contracted both tuberculosis and the herpes virus, squashing dreams that they could be released back into their forest homes.[43] Intestinal problems are less easily controlled because infestation with gastrointestinal parasites often occurs through soil eating.[44] There have also been reports of gastroenteritis and the occasionally fatal eosinophilic enterocolitis.[45] And the common cold, headache, and toothache can afflict orangutans just as much as humans. Gingivitis and periodontitis occur frequently in captivity,[46] and perinatal bacterial infections, diarrhea, and cardiac fibrosis also rank high among orangutan illnesses.[47]

Rehabilitation centers have learned a great deal from experience. Veterinary procedures are now in place at most centers to deal effectively with the health problems of orangutans taken into care and those leaving the centers.[48] Some, however, do not have veterinarians in attendance regularly. In Wanariset, veterinary care is extensive, and contact with humans (tourists and others) is kept to a minimum to avoid contagion.[49]

Health issues can pose substantial problems for translocation, reintroduction, or release of orangutans because illness either prevents the release of orangutans or, if undetected, could decimate the wild population after the release of just one infected animal. Release programs have been devised in some areas to translocate orangutans only into areas that are free of wild orangutan populations.[50] This avoids contamination, but it raises the question of why an area is free of wild orangutan populations. It may not be a sustaining environment, or it could be undesirable for other reasons. Weighing up the risks of release is a difficult exercise, particularly as there is rarely any information about the structure and composition of a projected release area.

The final problem of even the best and most integrated rehabilitation programs is where to release orangutans as the forest diminishes. Fragmentation of primary (virgin) and even of secondary (regrowth) forests poses one difficulty. This is because logging activity leaves pockets of forest that are too small to be self-sustaining and hence not useful for the orangutan. This is particularly so when one patch of forest is separated from the next by human habitation or plantations. Another difficulty is that one of the best methods of rehabilitation, that of "soft-release," will be impossible in the future because of the lack of hinterland near the rehabilitation stations. Soft-release means that the animal is released into an area near the former human carers so that it can obtain support when needed (usually food support). This method works for many native species in Australia[51] and elsewhere[52]—this form of release softens the transition and eases the animal into independence gently, giving it time to explore the environment without the full weight of responsibility for self-feeding and self-management. Gradually, the animal needs feeding assistance less and less and becomes a competent and independent individual without suffering debilitating and performance-reducing stress. It will move on and find its own area away from humans.

Sepilok Rehabilitation Centre was established for the soft-release of orangutans, but as demands grew, the center could not keep to its original plan. Like many other rehabilitation centers, Sepilok has become a target for ecotourism and is swamped by tourists. This creates substantial obstacles to its original aims. In addition, Sepilok was deemed too small an area to sustain the existing orangutan population in 1986, but continued to take in orangutans.[53] To our knowledge, there is no natural corridor to the remaining and relatively close areas that still had healthy orangutan populations some time ago, such as part of the hinterland of Sepilok and forests in the Sukau area across the bay along the Kinabatangan River. Where orangutans are not barred from traveling by sea or mangrove swamps, they are barred by human population and agriculture. Any new releases would now have to be far away from the center.

Despite all the very real problems faced by rehabilitation centers, they are still the greatest hope for the orangutans that have been forced into human company. At least at the centers, the orangutans are located in the climate and habitat where they belong, and some, like "our" Abbie, grow to adulthood with the kind of skills that may sustain them throughout adult life. Although small in numbers, individual success stories have been told. The Western world could do a great deal more to help these centers by sending money instead of tourists.

10
A FUTURE?

Is there not a way for orangutans to coexist with current plans for development in the rain forest? Some increased occupation of forest areas by humans will probably not matter, and for now, human occupation is almost the smallest ingredient in the equation. The forests are logged by those who have special licenses to do so, companies and individuals with considerable power. They want to make money, and there is a sizable short-term gain to be made. When Sabah had its timber export boom, there were more millionaires in the capital city, Sandakan, than in most large Western cities. Still, there are some restrictions at work now, and these go by the international catchwords of "sustainable" forestry and "selective logging." How do species living in the rain forest fare under this new regimen? Are these new forms of forestry the way of the future? The question is, whose future? Will they assure a place for the orangutan?

Cautious Destruction and the Orangutan

Let us explain what "sustainable" forestry and "selective logging" mean and how they impinge upon the life of the orangutan. Borneo still has enormous stretches of wilderness, but logging licenses have thrown a net over most of these vast areas. While many of them may still look like wilderness, they may be regrowth forests. On paper, they are already commercial crops.

To understand what selective logging and reforestation mean in a rain forest, we mention first a few of the characteristics of the rain forest that are particularly at risk from the new practices. Unlike common misconception, the soil of most old rain forest is extremely poor, and most nutrients are actually above ground. As a botanist explains: "The soil conditions [in the tropical rain forest] are only maintained through complex interaction of bacteria, insects, fungi and the function of root systems."[1] Reflecting these soil conditions, rates of growth in rain forests are excessively slow. Flowering and fruiting is irregular, even erratic, so that species that feed on fruits and flowers need large territories to survive, or they must migrate—the orangutan does both. Individual plant species are generally sparsely dispersed, but the forest compensates for this by sustaining an enormous variety of species.

One strategy that is widely employed in logging operations now is replanting, which is said to make forestry sustainable. Replanting guarantees that there is replacement and, in the best scenario, will include some of the species that have been felled. However, nowhere has it yet been possible to recreate the same variety and conditions of the forest as found before logging. Organic growth follows a number of sophisticated and complicated rules, and the rain forest is the most complex of all forests in its erratic, irregular, and diversified existence. Regrowth may be positive, but there are species that need 200 years to reach full height and at least sixty years to attain sexual maturity.

Current logging cycles operate in twenty-five-year or thirty-five-year intervals. This means that a tree is left to grow for twenty-five years or a little longer and then it is cut. The problem is that many trees have not reached sexual maturity by then and will have had no chance to reproduce. The logging cycle is not made with the preservation of the species in mind but with the minimum growth of a tree that makes its harvesting worthwhile. Also, it is generally the old and hence the best and tallest trees that are removed and replaced by new planting. The old trees and the tallest trees are the trees that an orangutan needs to move in. When they are removed, the orangutan is forced into the lower canopy. Moreover, the time gap between the tree that has been cut and the adequate growth of the one being planted may be 100–200 years. The gap between removal of forest and even partial replenishment can exceed several generations. With continued logging, the gap between clearance and regrowth can only widen. There are reforestation and afforestation programs in Borneo, but these tend to suffer from a lack of diversity and a possibly futile race against time. Forests are being cut at a faster (and accelerating) rate than new trees can grow.

Then there is the strategy of "selective logging," widely employed now in many countries. This means only a partial harvesting of timber in a plot, and in this way, the argument goes, not everything is disturbed. Roughly, only ten trees per hectare are harvested at any given time, which, indeed, sounds cautious and reasonable. However, the trees are not being lifted out by helicopters. Roads must cut through the forests to obtain the trees. Generally, part of the selective-logging procedure is to avoid clear-cutting whole patches of forest. Instead, the targets are individual trees sprinkled throughout the forest. Such selectivity means decreased accessibility. To get to the one tree, many other trees are torn down. For every ten trees harvested, another seventy are damaged or killed, amounting to about 40 percent of clearing in one particular forest patch.

Usually, the trees without value are not cut but left to rot or burn. They may not be good enough in quality or exactly what the buyer wants. The trees that are targeted are indeed logged "selectively," but the damage that is caused by this practice is extensive. The wastage in tons of timber is phenomenal, and the ruin left behind is a sad sight. In countries like Australia and Borneo and in Central and South America, there are now sites where the majority of the felled or damaged trees never get to market but are just left behind, destroyed.

In rain forests especially, selective logging affects a number of other variables, often with bad and even irreversible effects on the forest and its species. Removal of cover from even part of the forest floor creates problems of soil compaction, erosion, loss of nutrients, and increased exposure. Density of cover is the secret to the success of tropical rain forests. Take it away, and the forest will eventually succumb. We know that clearing affects humidity, and many species are humidity dependent for their germination. Loss of humidity through clearing thus alters the course of regeneration, slowing it down and leading to the successful germination of only a select number of species.[2] Plant diversity is thereby reduced and the extinction of plant species facilitated.

How Do Rain Forests and Orangutans Go Together?

Even an island like Borneo, not too long ago one of the last great wildernesses, is less and less able to claim possession of primeval virgin forest. Much of the forest now is regenerated forest. The problem with regrowth forests is not only their impoverished diversity but also the fact that they do not usually grow to their original height. This is a tremendously important factor for the orangutan. Unlike other forests, the rain forest is organized not in vertical layers but in three horizontal layers (canopy, undercover, and ground growth). Each layer has a living space that is measured horizontally, and animals living in the forest occupy their niches in layers. There are species that are predominantly upper-canopy dwellers, while others are middle-layer or ground dwellers. When the height of the forest is reduced, it affects the upper layer of the canopy most. Regenerating forests may show a reduction in height by as much as 25–50 percent. The animals occupying the upper canopy, therefore, suffer a significant loss of their horizontal living space.[3] As orangutans dwell in the canopy and are the largest mammal to be found in this layer of forest, they are affected most by logging. Even within fully regenerated forests, they lose the layer of canopy that they must occupy.

Loss in height of the canopy also affects natural regeneration. Rainforest trees and plants often depend on animals as agents of germination. Seeds are transported to new growth sites by animals, particularly the seeds of fruiting trees. Hornbills and orangutans fulfill this role either by consuming the seeds whole and disposing of them via feces or by eating the flesh and dropping the seeds. Hence, if orangutans are driven out of an area and replaced by less-demanding lower-canopy species such as squirrels, seed transport diminishes or changes. Squirrels are no replacement for the orangutans. They do not disperse seeds but digest them, so certain trees lose their germination agents and eventually disappear from the area.

Density of fruit is also affected in other ways. One study found, for instance, that adult and juvenile figs were less common in selectively logged forest than in primary forest.[4] The reason for this is that mature figs occur disproportionately in large trees and are particularly prevalent in the major commercial timber trees.[5] Therefore, even low-intensity selective logging drastically alters the density of fig trees and reduces the number of fruit-eating species, including hornbills *(Rhyticeros)* and orangutans *(Pongo pygmaeus)*.

Pearce might well have been right when he called sustainable forestry a "policy of wishful thinking."[6] There persists, even today, a sense of reassured calm that selective logging is a form of resource management based on sustainability. Most governments with high-density virgin rain forest have consented to map out reserves before an area is logged. These patches and island refuges for native plant and animal species are ultimately cosmetic procedures that will not halt, let alone reverse, the current rate of destruction. Without easy passage for animals from one protected region to the next, separated now by gulfs of human development, species diversity and genetic diversity within one species cannot be maintained.

There have been countless studies over the last two decades showing that reserves may only initially act as a refuge. When logging occurs, species flee to these areas, causing a sharp rise in animals, but within a six-month period, there is a sharp decline in the number of species found in the reserves.[7] In other words, while reserves may function as refuges from catastrophic events such as logging, they are generally too small to sustain the numbers of species for any length of time thereafter. Logging creates refugee animals with nowhere to go.

In an ironic twist, the destructive spiral of logging can eventually make animal species party to the destruction of the forest. When the lush forest has been turned into an ecologically fragile and confined environment, the original human and animal inhabitants suddenly become helpers in its destruction merely by continuing in the lifestyle that may have sustained them and the forest for untold centuries. The same species that once helped to maintain the forest can hasten the extinction of individual plant species when confined to areas that are too small for them. Orangutans wreak havoc on trees when they remain in the same area for too long. They take branches for nest building, play, and food and eventually strip trees completely. Spread over a large area, these habits result in tree pruning and have no detrimental effect, but in a small area, the devastation quickly becomes visible, as is the case in Sepilok. This destruction in turn aids their own demise because it limits sustainable living space even further. Most undisturbed forest reserves are already too small to be self-sustaining, meaning that the fauna in them are reduced and not self-perpetuating. Some of the nature reserves may be extremely small, others are 50–100 hectares or larger. Sepilok has 10,000 hectares.

Natural and Man-Made Disasters

In 1997 and 1998 further tragedy struck Indonesian Borneo and Sarawak. The international press reported that the forests of Borneo were burning. The fires in January 1998 were particularly devastating. They alone destroyed over 5 million hectares of orangutan habitat. When friends of ours were staying in Balikpapan in June, there were still nearly forty fires burning near the town. One little fire truck, an ancient model, was not even taken out of its carport because it would have been useless against the magnitude of the problem.

It is believed that some of the fires were intentionally lit. There are plans for the relocation of people to Borneo from overcrowded Jakarta. Rice fields must replace jungle. There are private companies that wish to utilize the land for monoculture, so the forest has to go. Companies setting up plantations prefer to clear forest rather than utilize the ever-increasing areas of grasslands because they can use the timber for building; and when planting, they benefit from the better soil conditions due to the litter covering of the forest floor. Except for the indigenous people, for whom the forest is the main food supply, there are many parties at national and multinational levels that can benefit from a retreat of the forest.

International reaction to the fires was confused and lukewarm. There were no strong cries of "save the rain forest" that might have resulted in the kind of international financial support needed at the time to stop the fires. One suspects that this was partly so because interest in putting out the fires by those in power and those with vested interests was ambivalent or nonexistent.

Small farmers have always burnt off small patches of forest (different patches each year) during the dry season, but in the 1997–1998 period the forest was tinder-dry. There had been little or no rain. It was hot. The rains had not come, so the fires got out of hand. Some fires were meant to do that, but others were not. For months on end, the air of the entire Bornean island was filled with heavy smoke, so heavy, in fact, that the airport at Kuching had to be closed for some time because the smoke was five times the World Health Organization's standards of air pollution. Friends of ours returned from Sandakan in Sabah, the other Eastern Malaysian state, and warned us not to go. They had to cut their visit short because the smoke became too much to bear. The smoke alone was more than a health risk. It was a killer to animals and humans alike.

Primeval forest is usually too moist to burn out of control, but secondary and regrowth forest has lost some of its water-retention ability, so it dries out sufficiently to become vulnerable to fire. Unlike Australian eucalyptus, Bornean rain-forest plants do not benefit from the fire, nor do species recover, except for the *Anthocephalus cadamba* trees, which can withstand fire and drought. Most regrowth forest plants just burn and die. Unless they can fly, all the animals within the forest die, too. Tropical forest species do not expect fires. They are not programmed to deal with them as, to some extent, animals are in regions with regular fires (e.g., species of the African savannah or parts of the Australian bush). For Bornean forest species, there is nowhere to run. Willie Smits, of the research station in Wanariset, filed this report in his diary on August 4:

> My friend Deddy who works in the concession recalls the terrible situation when the fires were raging . . . [there was] a large adult male orangutan in the top of a dead tree isolated in the middle of a sea of fire. Deddy could see him from fifty meters distance from a higher slope but was not able to do anything. The orangutan male repeatedly threw his head backwards and cried enormously loud while tears were running over his cheekpads. When the fire reached his legs he dropped down forty meters into the flames. This intelligent orangutan must have realised from the moment he got isolated what would happen to him.[8]

In another entry (August 10), he mentions seeing a very lonely orangutan in an acacia tree and is overcome with sadness by the tragedy that has occurred:

> The rain trickles down, the sky is grey, a chainsaw roars in the distance and the sight of this weak animal hanging/sleeping at this time of day from a branch in one of the last five trees still standing in the barren landscape truly confers the tragedy we are witnessing.[9]

But this was not all. Once the fires had gone and the media had lost interest in the situation of Kalimantan, more disaster was yet to come, barely, if at all, reported in the world news. The punishments descending on the areas affected by fire were truly of biblical proportions. The fires had just stopped when the rains came, spurred on by La Niña, an unusual amount of rain that flooded whole areas. Towns stood up to two meters underwater. In Samarinda, a dam broke and put 30,000 houses underwater. Dead trees, killed by the fire, stood in deep water. The fields were underwater, and the much-needed harvest failed. The people could not eat. They had lost their few belongings, and by August 1998, hunger was everywhere. Rice huts and transmigration villages (for seasonal workers) lay abandoned and falling apart.

After the fire, the floods finished the destruction of the rain forest in those areas forever. The fires had led to deterioration of the soil structure and loss of the water-holding litter layer, and now the water washed away nutrients such as the ashes from the fire and the remaining leaves and roots. Rivers become choked with these nutrients, which did great damage to animal life in the water. And finally, the water washed away the soil itself, furrowing deep into the ground and taking with it what it could. There were no obstacles to stop the water, and the floods rose much faster than they would have done had the forest still been there.

By August 1998, the full aftermath of the fires and floods was just becoming clear. Underneath the burned trees is a hue of green, not a signal of regrowth but the first sign of the final end of the forest. It is the growth of the infamous alang-alang *(Imperata cylindrica)* grasses. These grasses not only increase the risk of future fires but actively

FIGURE 10.1 *Increasingly, even the faces of orangutans in their natural habitat look sad. This is not surprising, given the many traumas they have recently endured.*

inhibit the establishment of trees. Their roots excrete substances akin to cyankali, which affects the germination of many tree seeds.[10]

In the aftermath, hunger, starvation and disease set in, for human populations as well as for orangutans (Figure 10.1). Wanariset Station was holding 171 refugee and injured orangutans in August 1998, most of them extremely young. Mothers had died of starvation trying to sustain themselves on acacia bark, and their young ones were so weak they were unable to hold on any longer and just fell to the ground. Some mothers were eaten by the starving local population, others succumbed to disease.

Where to Next?

The orangutans that have survived face an uncertain future. No one knows exactly how many orangutans died as a result of the droughts, the fires, the smoke, and the floods, but we know that these events

FIGURE 10.2 *Last vestiges of paradise—a tributary of the Kinabatangan River.*

have had a devastating and lasting effect on orangutans and may just have been enough to set the final scene for their extinction in Borneo in the not-too-distant future.

In 1995, well before these events, Mark Leighton from Harvard University and colleagues from habitat countries and wildlife groups evaluated the survival chances of orangutans. Calculating the interactions of infant mortality, adult mortality, and interbirth intervals (on average eight years), they concluded that the wild populations could not sustain an increase in adult female mortality. They found that the removal of a mere five adults per 1,000 orangutans per year would lead to a long-term decline of the population. Further removal of adult female orangutans (as a consequence of the pet trade and local conditions, as described above) would increase the rate of decline and the risk of extinction.[11]

This was before the fires. Catastrophic events happen, but we expect recovery to follow. For the orangutans, and possibly for many other species, it would be wrong and irresponsible to promote a glib optimism. Recovery for the orangutans is only possible if their habitat is not at risk. The fires have burned their habitat, and it is not going to return. Thousands of orangutans lost their ancient home in a single year.

We asked in our previous book, *Orangutans in Borneo*: "What has the orangutan done to us?" Nothing, nothing at all. In the late 1990s, free-ranging orangutans are suffering. They are hounded and hunted and pushed and shoved at the very time when international discourse talks about rescue missions, recovery, rehabilitation, protective legislation, and humane and conservationist goals.

Everything is in place—but only on paper. There are often individual actions of near heroic levels, but the international will is distorted and dampened by complex politics and conflicting plans in an economic and political global context that is only too ready to mask indifference with strategic plans and reassuring conservationist-sounding goals.

The reality and hope for Bornean orangutans lies only in their last remaining patches of rain forest, heavily contested as they are (Figure 10.2). In our opinion, orangutans are too large to keep in enclosures, too intelligent to keep in zoos, too self-aware to keep in laboratories, too sensitive to be exploited in shows and circuses, and too close to us to ignore the fact that they too have a right to live freely.

NOTES

Chapter 1

1. MacKinnon 1971.
2. For details, see Kaplan and Rogers 1994; Rogers 1998.
3. See Kahlke 1972; von Koenigswald 1982.
4. Harrisson 1957; Ho 1990; Hooijer 1948, 1960.
5. Smith and Pilbeam 1980.
6. Sugardjito and van Hooff 1986.
7. MacKinnon 1971, 1974.
8. Whitmore 1987, and Figure 8 in von Koenigswald 1982.
9. Hall 1997.
10. Peltier 1994.
11. See web site by Muir, Watkins, and Ha 1998.
12. See van Oosterzee 1997.
13. Tilson et al. 1993.
14. Groves, Westwood, and Shae 1992.
15. See de Boer 1982, and Chapter 2.
16. Tilson et al. 1993.

Chapter 2

1. See Jones, Martin, and Pilbeam 1992 for more details.
2. Muir, Galdikas, and Beckenbach 1995.
3. Gribbin and Gribbin 1988.
4. Easteal and Herbert 1997.

5. Discussed in Chapter 1; see also Groves 1989.
6. Arnason et al. 1996.
7. Penny, Murray-McIntosh, and Hendy 1998.
8. Lewin 1984; Wolpoff 1989.
9. Avise 1991.
10. See preface to Nadler et al., *The Neglected Ape* (New York: Plenum Press, 1995).
11. Schwartz 1984, 1987.
12. Weiss 1987.
13. Mohammad-Ali, Eladari, and Galibert 1995.
14. For more detail, see Spuhler 1988.
15. See Bauer and Schreiber 1995 for a different method giving the same result.
16. Zhi et al. 1996.
17. Xu and Arnason 1996.
18. Muir, Galdikas, and Beckenbach 1998.
19. E.g., blue whales and fin whales, as Xu and Arnason 1996 point out; also, Arnason 1998.
20. Galdikas 1995.
21. Rogers and Kaplan 1998.
22. Lewin 1988.
23. See Cavalieri and Singer 1993; Rogers 1997.

Chapter 3

1. Davenport 1967.
2. Veevers-Carter 1991.
3. Utami et al. 1997.
4. Galdikas 1984.
5. Western 1994.
6. Ungar 1995.
7. Hamilton and Galdikas 1994.
8. Huffman and Seifu 1989; Takasaki and Hunt 1987.
9. Dossaji, Wrangham, and Rodriguez 1989.
10. Köhler 1921; Harrisson 1960.
11. Utami and van Hooff 1997.
12. Rodman 1977.
13. Sugardjito and Nurhuda 1981.
14. Boesch 1990.
15. Stanford et al. 1993.
16. Schürman and van Hooff 1986.
17. Sugardjito, te Boekhorst, and van Hooff 1987.

18. Mitani et al. 1991.
19. See Normile 1998.
20. Galdikas 1984.
21. Zuckerman 1981.
22. Rodman and Mitani 1987.
23. Nadler 1988; van Hooff 1995.
24. Hornaday 1885.
25. Leighton et al. 1995.
26. te Boekhorst, Schürmann, and Sugardjito 1990.
27. Falk 1998.
28. Falk 1992.
29. Semendeferi et al. 1997.
30. Rilling and Insel 1998.
31. Geschwind and Levitsky 1968; also summarized in Bradshaw and Rogers 1993.
32. Le May and Geschwind 1975.
33. Rogers and Kaplan 1996.

Chapter 4

1. Galdikas 1981.
2. Asano 1967; Lippert 1974; Markham 1990.
3. Galdikas 1982.
4. Coe 1990.
5. Galdikas 1982.
6. Galdikas and Wood 1990.
7. Baldwin 1986.
8. Codner and Nadler 1984.
9. Vochteloo et al. 1993.
10. Anne Russon, personal communication, 1998.
11. Nadler 1983.
12. Shively and Mitchell 1986.
13. Maple and Hoff 1982; Shively and Mitchell 1986.
14. Coe 1990.
15. Davenport 1967.
16. Harrisson 1960.
17. Baldwin 1986.
18. Horr 1977.
19. Maple and Zucker 1978.
20. MacKinnon 1971.
21. James 1890.

22. Lewis and Coates 1980.
23. Bard et al. 1992.
24. Ibid.
25. Horwich 1989.
26. Davenport 1967.
27. MacKinnon 1971.
28. Davenport 1967; MacKinnon 1971.
29. Davenport 1967.
30. Lewis 1978.
31. Rijksen and Rijksen-Graatsma 1975; Aveling and Mitchell 1982.

Chapter 5

1. Byrne 1995.
2. King 1994.
3. Shapiro 1985.
4. Shapiro and Galdikas 1995.
5. Rogers and Kaplan 1998.
6. Kaplan and Rogers 1996a, 1996b.
7. Meltzoff 1996.
8. Moore 1996.
9. Russon and Galdikas 1995a.
10. Byrne 1997; Whiten and Ham 1992.
11. Visalberghi and Fragaszy 1990; Visalberghi and Limongelli 1994.
12. Byrne 1988.
13. Boesch 1993; Matsuzawa 1994, 1996.
14. Bard 1993.
15. Rijksen 1978; Galdikas 1982.
16. Russon and Galdikas 1995a.
17. Galdikas 1982; Russon and Galdikas 1993.
18. Russon and Galdikas 1993.
19. Tomasello 1993.
20. MacKinnon 1971.
21. Davenport 1967.
22. Bard 1993.
23. Thorpe 1963.
24. Peters 1995.
25. Russon and Galdikas 1995b.
26. Ibid.
27. Boesch 1991, 1993.
28. Corner 1949.

29. Veevers-Carter 1991:91.
30. Parker 1996.
31. Pagnel and Harvey 1993.
32. Pereira and Altmann 1985.
33. Harrisson 1960.
34. Kaplan and Rogers 1999.
35. Galdikas 1982.

Chapter 6

1. Walker 1964.
2. Rogers and Kaplan 1998a.
3. Lethmate 1982.
4. Parker 1969.
5. Visalberghi, Fragazy, and Savage-Rumbaugh 1995.
6. Lethmate 1982.
7. van Schaik and Fox 1996.
8. Ibid.
9. MacKinnon 1974.
10. Schaller 1961.
11. Rogers and Kaplan 1994.
12. Bard 1990.
13. See Rogers 1997 for further details.
14. Cant 1987.
15. Povinelli and Cant 1995.
16. Galdikas 1982.
17. See Rogers 1997.
18. Miles 1990.
19. Povinelli and Preuss 1995.
20. Tobach et al. 1997.
21. Povinelli and Cant 1995.
22. Hauser et al. 1995.
23. Summarized in Rogers 1997.
24. Call and Tomasello 1995.
25. Call and Tomasello 1994.
26. Itakura and Tanaka 1998.
27. Tomasello, Call, and Gluckman 1997.
28. Call and Rochat 1996.
29. Call and Rochat 1997.

Chapter 7

1. Poole 1987; Zucker and Thibaut 1995.
2. Masson and McCarthy 1995.
3. Rogers and Kaplan 1998.
4. Miles 1990; Bard 1992; Call and Tomasello 1994.
5. Smits 1998.
6. MacKinnon 1971.
7. Marler and Tenaza 1977.
8. Tomasello, Gust, and Frost 1989.
9. Bard 1992.
10. Bard 1993.
11. Chevalier-Skolnikoff 1974.
12. Ekman 1974.
13. van Hooff 1972.
14. Redican 1975.
15. MacKinnon 1974; Rijksen 1978; Maple 1980.
16. van Hooff 1972.
17. Ibid.
18. Provine 1996.
19. Gautier and Gautier 1977.
20. Yamagiwa 1992.
21. Bard 1990.
22. Miles 1990; Call and Tomasello 1994.
23. Rogers 1997.
24. Chance and Jolly 1970.
25. Kaplan and Rogers 1998.
26. Rogers and Kaplan 1998.
27. MacKinnon 1971; Niemitz and Kok 1976.
28. Kaplan and Rogers 1994.
29. Galdikas and Insley 1988.
30. Kawabe 1970.
31. MacKinnon 1971; Mitani 1985.
32. MacKinnon 1971.
33. Galdikas and Insley 1988.
34. Ibid.
35. Cheney and Wrangham 1987.
36. Bradshaw and Rogers 1993.
37. Greenfield and Savage-Rumbaugh 1990.
38. Shapiro and Galdikas 1995.
39. Miles 1990.
40. Shapiro 1982.

41. Miles 1994.
42. Miles 1990.
43. Rogers 1997.

Chapter 8

1. Voigt 1984.
2. Harrisson 1960.
3. Rijksen 1978.
4. MacKinnon 1971.
5. Collins, Graham, and Preedy 1975; Graham 1981a.
6. Leighton et al. 1995.
7. Masters and Markham 1991.
8. Winkler 1989.
9. Maple 1980.
10. Kingsley 1982.
11. Winkler 1989.
12. Galdikas 1985b.
13. Ibid.
14. Ibid.
15. Kavanagh 1983.
16. Galdikas 1995.
17. Kavanagh 1983.
18. Schwartz 1984.
19. Horr 1975.
20. Ibid.
21. Hornaday 1885; Galdikas 1985.
22. Galdikas 1985.
23. Graham 1981b.
24. Kavanagh 1983; also our own observation.
25. Kavanagh 1983.
26. Thompson-Handler, Malenky, and Badrian 1984.
27. Kavanagh 1983.
28. Galdikas and Wood 1990.
29. MacKinnon 1971.
30. Rodman and Mitani 1987.
31. Graham and Nadler 1990.
32. Sodaro 1988.
33. Vandevoort et al. 1993.
34. Ibid.
35. Asa et al. 1995.

36. Nadler 1995.
37. Vandevoort et al. 1993.
38. Asa et al. 1995.
39. Perkins and Maple 1990.
40. Cocks 1998.
41. Crawford and Galdikas 1986.
42. Markham 1990.
43. Cantor 1993.
44. Ibid.

Chapter 9

1. Booth 1988.
2. Kaur 1998.
3. Yerkes and Yerkes 1945.
4. Ibid.
5. Reynolds 1967.
6. Yerkes and Yerkes 1929.
7. Köhler 1921.
8. Kaplan and Rogers 1995.
9. Tarling 1989.
10. Williams 1991.
11. Eudy 1994.
12. Ibid.
13. Phipps 1992.
14. Cantor 1993.
15. Schlentrich and Ng 1994.
16. Briggs 1991.
17. McNeely and Thorsell 1989.
18. *Northern Daily Leader* 1997.
19. Winch 1996.
20. Burns 1994.
21. Kaplan 1998.
22. Raven 1988; Whitmore and Sayer 1992.
23. Pearce 1990.
24. MacKinnon 1987.
25. IFPNC 1993.
26. Ibid.
27. Black 1998.
28. Dobson 1996.
29. *International Zoo Year Book 1959–1965*.

30. Perkins 1994.
31. Cocks 1998.
32. See Dobson 1996.
33. Masson and McCarthy 1995:53.
34. Masson and McCarthy 1995.
35. Shepherdson 1994.
36. Lardeux-Gilloux 1995.
37. Schaller 1965.
38. Discussed further by Rogers and Kaplan 1998.
39. Cocks 1998.
40. Francis 1958; Fiennes 1979.
41. Jones 1982.
42. Kehoe, Phin, and Chu 1984.
43. Woodford and Rossiter 1994.
44. Munson and Montali 1990.
45. Citino et al. 1985.
46. Braswell 1994.
47. Munson and Montali 1990.
48. Sayuthi et al. 1991.
49. Smits, Herihanto, and Ramono 1995.
50. Peters 1995.
51. Kaplan 1999.
52. Wildlife Rehabilitation Statistics 1998.
53. Davies 1986.

Chapter 10

1. Veevers-Carter 1991.
2. Ng 1983.
3. Ibid.
4. Leighton and Leighton 1983.
5. Ibid.
6. Pearce 1990.
7. Lovejoy et al. 1983.
8. Smits 1998.
9. Ibid.
10. Ibid.
11. Leighton et al. 1995.

REFERENCES

Arnason, U. (1998) Response to criticism by C. Muir et al. *Journal of Molecular Evolution*, 46, 379–81.

Arnason, U., Gullberg, A., Janke, A. and Xu, X. (1996) Pattern and timing of evolutionary divergences among hominoids based on analysis of complete mtDNAs. *Journal of Molecular Evolution*, 43, 650–61.

Asa, C.S., Silber, S.J., Porton, I., Fischer, F., Lenehan, K., Deters, M., Junge, R., Hicks, J. and Cohen, R.C. (1995) Follicle stimulation and ovum collection in the orangutan. In R.D. Nadler, B.F.M. Galdikas, L.K.Sheeran and N. Rosen (eds.), *The Neglected Ape*. Plenum Press, New York, 217–22.

Asano, M. (1967) A note on the birth and rearing of an orang-utan *Pongo pygmaeus* at Tama Zoo, Tokyo. *International Zoo Yearbook*, 7, 95–6.

Aveling, R.J. and Mitchell, A. (1982) Is rehabilitation of orang-utans worthwhile? *Oryx*, 16, 263–71.

Avise, J.C. (1991) Matriarchal liberation. *Nature*, 352, 192.

Baldwin, J.D. (1986) Behaviour in infancy: exploration and play. In G. Mitchell and J.Erwin (eds.), *Behavior, Conservation, and Ecology*. Alan R. Liss, New York. 2, partA: 295–326.

Bard, K.A. (1990) "Social tool use" by free-ranging orang-utans: a Piagetian and development perspective on the manipulation of an inanimate object. In S.T. Parker and K.R. Gibson (eds.), *"Language" and Intelligence in Monkeys and Apes: Comparative Developmental Perspectives*. Cambridge University Press, Cambridge, 356–78.

______. (1992) Intentional behaviour and intentional communication in young free-ranging orang-utans. *Child Development*, 63, 1186–97.

______. (1993) Cognitive competence underlying tool use in free-ranging orang-utans. In A. Berthelet and Chavaillon (eds.), *The Use of Tools by Human and Non-human Primates*. Oxford University Press, Oxford, 103–17.

Bard, K.A., Platzman, K.A., Lester, B.M. and Suomi, S.J. (1992) Orientation to social and nonsocial stimuli in neonatal chimpanzees and humans. *Infant Behavior and Development*, 15, 43–56.

Bauer, K. and Schreiber, A. (1995) Tricky relatives: consecutive dichotomous speciation of gorilla, chimpanzee and hominids testified by immunological determinants. *Naturwissenschaften*, 82, 517–20.

Black, J.M. (1998) Threatened waterfowl: recovery priorities and reintroduction potential with special reference to the Hawaiian Goose. In J.M.Marzluff and R. Sallabanks (eds.), *Avian Conservation. Research and Management*. Island Press, Washington, D.C., 125–40.

Boesch, C. (1990) First hunter of the forest. *New Scientist*, 19 May, 20–5.

______. (1991) Teaching among wild chimpanzees. *Animal Behaviour*, 41, 530–2.

______. (1993) Aspects of transmission of tool-use in wild chimpanzees. In K.R.Gibson and T. Ingold (eds.), *Tools, Language and Cognition in Human Evolution*. Cambridge University Press, Cambridge.

Booth, W. (1988) Reintroducing a political animal. *Science*, 24, 156–9.

Bradshaw, J.L. and Rogers, L.J. (1993) *The Evolution of Lateral Asymmetries, Language, Tool Use, and Intellect*. Academic Press, San Diego, CA.

Braswell, L.D. (1994) *Oral findings in captive orangutans*. International orangutan conference: The Neglected Ape, Fullerton, California State University.

Briggs, J. (1991) *Parks of Malaysia*. Longman Malaysia, Petaling Jaya, Selangor.

Burns, J. (1994) Go ape over orang-utans! *Sunday Telegraph*. Sydney, 30 October, 146.

Byrne, R.W. (1995) *The Thinking Ape: Evolutionary Origins of Intelligence*. Oxford University Press, Oxford.

______. (1997) The technical intelligence hypothesis: an additional evolutionary stimulus to intelligence? In A. Whiten and R.W. Byrne (eds.), *Machiavellian Intelligence II. Extensions and Evaluations*. Cambridge University Press, Cambridge, 289–311.

Byrne, R.W. and Whiten, A. (1988) *Machiavellian Intelligence: Social Enterprise and the Evolution of Intellect in Monkeys, Apes and Humans*. Clarendon Press, Oxford.

Call, J. and Rochat, P. (1996) Liquid conservation in orangutans (*Pongo pygmaeus*) and humans (*Homo sapiens*): individual differences and perceptual strategies. *Journal of Comparative Psychology*, 110, 219–32.

______. (1997) Perceptual strategies in the estimation of physical quantities by orangutans (*Pongo pygmaeus*). *Journal of Comparative Psychology*, 111, 315–29.

Call, J. and Tomasello, M. (1994) Production and comprehension of referential pointing by orangutans (*Pongo pygmaeus*). *Journal of Comparative Psychology*, 108(4), 307–17.

______. (1995) Use of social information in the problem solving of orang-utans *(Pongo pygmaeus)* and human children (*Homo sapiens*). *Journal of Comparative Psychology*, 109, 308–20.

Cant, J.G.H. (1987) Positional behavior of female Bornean orang-utans (*Pongo pygmaeus*). *American Journal of Primatology*, 12, 71–90.

Cantor, D. (1993) Items of property. In P. Cavalieri and P. Singer (eds.), *The Great Ape Project. Equality beyond Humanity*. Fourth Estate, London, 280–90.

Cavalieri, P. and Singer, P. (eds.) (1993) *The Great Ape Project: Equality beyond Humanity*. Fourth Estate, London.

Chance, M.R.A. and Jolly, C.J. (1970) *Social Groups of Monkeys, Apes and Man*. Jonathan Cape, London.

Cheney, D.L. and Wrangham, R.W. (1987) Predation. In B.B. Smuts, D.L. Cheney, R.M. Seyfarth, R.W. Wrangham and T.T. Struhsaker (eds.), *Primate Societies*. Chicago University Press, Chicago, 227–39.

Chevalier-Skolnikoff, S. (1974) Facial expression of emotion in nonhuman primates. In P. Ekman (ed.), *Darwin and Facial Expression*. Academic Press, New York.

Citino, S.B., Bush, M., Phillips, L.G., et al. (1985) Eosinophilic enterocolitis in a juvenile orangutan pongo hybrid. *Journal of the American Veterinary Medical Association*, 187(11), 1279–80.

Cocks, L. (1998) *Factors Influencing the Well-being and Longevity of Captive Female Orang-utans*. Primatology Conference, Kuching, Sarawak and Perth Zoo.

Codner, M.A. and Nadler, R.D. (1984) Mother-infant separation and reunion in the great apes. *Primates*, 25(2), 204–17.

Coe, C. (1990) Psychobiology of maternal behavior in nonhuman primates. In N.A. Krasnegor and R.S. Bridges (eds.), *Mammalian Parenting, Biochemical, Neurobiological, and Behavioral Determinants*. Oxford University Press, New York and Oxford, 157–83.

Collins, D.C., Graham, C.E. and Preedy, J.R.K. (1975) Identification and measurement of urinary estrone, estradiolB17ß, estriol, pregnanediol and androsterone during the menstrual cycle of the orang-utan. *Endocrinology*, 96, 93–101.

Corner, E.J.H. (1949) The durian theory and the origin of the modern tree. *Annals of Botany NS*, XVII(52).

Crawford, C. and Galdikas, B.M.F. (1986) Rape in non-human animals: an evolutionary perspective. *Canadian Psychology*, 27(3), 215–30.

Davenport, R.K. (1967) The orangutan in Sabah. *Folia Primatologia*, 5, 247–63.

Davies, G. (1986) The orangutan pongo-pygmaeus in Sabah. *Oryx*, 20(1), 44–5.

de Boer, L.E.M. (1982) Genetics and conservation of the orang-utan. In L.E.M. de Boer (ed.), *The Orang Utan: Its Biology and Conservation*. Dr W. Junk Publishers, The Hague, 39–60.

Dobson, A.P. (1996) *Conservation and Biodiversity*. Scientific American Library, New York.

Dossaji, S.F., Wrangham, R. and Rodriguez, E. (1989) Selection of plants with medicinal properties by wild chimpanzees. *Fitoterapia*, 60, 378–81.

Easteal, S. and Herbert, G. (1997) Molecular evidence from the nuclear genome for the time frame of human evolution. *Journal of Molecular Evolution*, 44, S121–S132.

Ekman, P. (ed.) (1974) *Darwin and Facial Expression*. Academic Press, New York.

Eudy, A.A. (1994) The impact of socioeconomic decisions on the status of the orang-utan and other East Asian fauna. In R.D. Nader, B. Galdikas, L. Sheehan and N.Rosen (eds.), *The Neglected Ape*. Plenum Press, New York, 23–7.

Falk, D. (1992) Evolution of the brain and cognition in hominids. *The Sixty-second James Arthur Lecture on the Evolution of the Human Brain*. The American Museum of Natural History, New York.

______. (1998) Hominid brain evolution: looks can be deceiving. *Science*, 280, 1714.

Fiennes, R. (1979) *Zoonoses of Primates*. Weidenfeld and Nicholson, London.

Finlay, B.L. and Darlington, R.B. (1995) Linked regularities in the development and evolution of mammalian brains. *Science*, 268, 1578–84.

Francis, J. (1958) *Tuberculosis in Animals and Man*. Cassell and Co, London.

Galdikas, B.M.F. (1981) Orang utan reproduction in the wild. In C. E. Graham (ed.), *Reproductive Biology of the Great Apes: Biomedical and Comparative Aspects*. Academic Press, New York, 281–300.

______. (1982a) Orang-utan tool-use at Tanjung Puting Reserve, Central Indonesian Borneo (Kalimantan Tengah). *Journal of Human Evolution*, 10, 19–33.

______. (1982b) Wild orangutan birth at Tanjung Puting Reserve. *Primates*, 23, 500–10.

______. (1984) Adult female sociality among wild orangutans at Tajung Puting Reserve. In M. Small (ed.), *Female Primates: Studies by Women Primatologists*. Alan R. Liss, New York, 217–35.

______. (1985a) Adult male sociality and reproductive tactics among orangutans at Tanjung Puting Borneo Indonesia. *Folia Primatologica*, 45(1), 9–24.

______. (1985b) Subadult male orangutan sociality and reproductive behavior at Tanjung Puting. *American Journal of Primatology*, 8(2), 87–100.

______. (1995a) *Reflections of Eden: My Life with the Orang-utans of Borneo*. Victor Gollancz, London.

______. (1995b) Social and reproductive behavior of wild adolescent female orang-utans. In R.D. Nadler, B.F.M. Galdikas, L.K. Sheeran and N. Rosen (eds.), *The Neglected Ape*. Plenum Press, New York, 163–82.

Galdikas, B.M. and Insley, S.J. (1988) The fast call of the adult male orangutan. *Journal of Mammalogy*, 69(2), 371–5.

Galdikas, B.M.F. and Wood, J.W. (1990) Birth spacing patterns in humans and apes. *American Journal of Physical Anthropology*, 83(2), 185–92.

Gautier, J.P. and Gautier, A. (1977) Communication in Old World monkeys. In T.A.Sebeok (ed.), *How Animals Communicate*. Indiana University Press, Bloomington.

Geschwind, N. and Levitsky, W. (1968) Human brain: left-right asymmetry in temporal speech region. *Science*, 161, 319–58.

Graham, C.E. (1981a) Menstrual cycle of the great apes. In C.E. Graham (ed.), *Reproductive Biology of the Great Apes: Biomedical and Comparative Aspects*. Academic Press, New York, 1–43.

______. (ed.) (1981b) *Reproductive Biology of the Great Apes: Biomedical and Comparative Aspects*. Academic Press, New York.

Graham, C.E. and Nadler, R.D. (1990) Socioendocrine interactions in great ape reproduction: socioendocrinology of primate reproduction. In T. E. Ziegler and F.B.Bercovitch (eds.), *Symposium from the XII Congress of the International Primatological Society*. International Primatological Society, Brasilia, Brazil, 33–58.

Greenfield, P.M. and Savage-Rumbaugh, E.S. (1990) Grammatical combination in *Pan paniscus*: processes of learning and invention in the evolution and development of language. In K.R. Gibson and S.T. Parker (eds.), *"Language" and Intelligence in Monkeys and Apes: Comparative Developmental Perspectives*. Cambridge University Press, 540–78.

Gribbin, J. and Gribbin, M. (1988) *The One Percent Advantage: The Sociobiology of Being Human*. Blackwell, Oxford.

Groves, C.P. (1989) *A Theory of Human and Primate Evolution*. Clarendon Press, Oxford.

Groves, C.P., Westwood, C.B. and Shae, B.T. (1992) Unfinished business: Mahalanobis and a clockwork orang. *Journal of Human Evolution*, 22, 327–40.

Hall, R. (1997) *Plate tectonics of Cenozoic SE Asia*. Animated diagrams available on http://glsun2.gl.rhbnc.ac.uk/seasia/welcome.html

Hamilton, R.A. and Galdikas, B.M.F. (1994) A preliminary study of food selection by the orangutan in relation to plant quality. *Primates*, 35, 255–63.

Harrisson, B. (1960) A study of orang-utan behaviour in semi-wild state, 1956–1960. *The Sarawak Museum Journal*, IX(15–16 [New Series]), 422–47.

Harrisson, T. (1957) The great cave of Niah: a preliminary report on Bornean prehistory. *Man*, 57, 161–6.

Hauser, M.D., Kralik, J., Botto-Mahan, C., Garrett, M. and Oser, J. (1995) Self-recognition in primates: phylogeny and the salience of species-typical features. *Proceedings of the National Academy of Sciences*, 92, 10811–14.

Ho, C.K. (1990) A new Pliocene hominoid skull from Yuanmou southwest China. *Human Evolution*, 5, 309–18.

Hooijer, D.A. (1948) Prehistoric teeth of man and of the orang-utan from central Sumatra, with notes on the fossil orang-utan from Java and southern China. *Zoologische Mededelingen*, 29, 175–301.

______. (1960) The orang-utan in Niah cave prehistory. *The Sarawak Museum Journal*, 9, 408–21.

Hornaday, W.T. (1885) *Two Years in the Jungle. The Experiences of a Hunter and Naturalist in India, Ceylon, the Malay Peninsula, and Borneo.* K. Paul, Trench and Co., London.

Horr, D.A. (1975) The Borneo orang-utan: population structure and dynamics in relationship to ecology and reproductive strategy. In L.A. Rosenblum (ed.), *Primate Behavior. Developments in Field and Laboratory Research.* Academic Press, New York, San Francisco, London. 4: 307–23.

______. (1977) Orang-utan maturation: growing up in a female world. In S. Chevalier-Skolnikoff and F.E. Poirier (ed.), *Primate Biosocial Development.* Garland, New York.

Horwich, R.H. (1989) Cyclic development of contact behavior in apes and humans. *Primates*, 30, 269–79.

Huffman, M.A. and Seifu, M. (1989) Observation on the illness and consumption of a possibly medicinal plant *Veronia amygdalina* (DEL) by a wild chimpanzee in the Mahale Mountains National Park, Tanzania. *Primates*, 30, 51–63.

IFPNC, Indonesian Forest Protection and Conservation and IUCN/SSC Captive Breeding Specialist Group. (1993) *Orangutan: Population and Habitat Viability Analysis Report.* Population and Habitat Viability Analysis Workshop for Orangutan, Medan, North Sumatra.

International Zoo Year Book, 1959–1965, vols. I–V.

Itakura, S. and Tanaka, M. (1998) Use of experimenter-given cues during object-choice tasks by chimpanzees (*Pan troglodytes*), an orang-utan (*Pongo pygmaeus*) and human infants (*Homo sapiens*). *Journal of Comparative Psychology*, 112, 119–26.

James, W. (1890) *The Principles of Psychology* vols. 1 and 2, Dover Publications, New York.

Jones, D.M. (1982) Conservation in relation to animal disease in Africa and Asia. *Symposium of the Zoological Society London*, 50, 271–85.

Jones, S., Martin, R. and Pilbeam, D. (eds.) (1992) *The Cambridge Encyclopedia of Human Evolution.* Cambridge University Press, Cambridge.

Kahlke, H.D. (1972) A review of the Pleistocene history of the orang-utan (*Pongo* Lacépède 1799). *Asian Perspectives*, 15, 5–14.

Kaplan, G. (1998) Economic development and Ecotourism in Malaysia. In A.Kaur and I. Metcalfe (eds.), *The Shaping of Malaysia.* Macmillan, London, 173–95.

______. (1999) Animal rehabilitation: Can exercise in the practice of biodiversity and a tool for conservation. *Animal Issues*, 3(1), 1–25.

Kaplan, G. and Rogers, L.J. (1994) *Orang-utans in Borneo.* University of New England Press, Armidale, NSW.

______. (1995) Of human fear and indifference: the plight of the orang-utan. In R.D. Nader, B. Galdikas, L. Sheehan and N. Rosen (eds.), *The Neglected Ape.* Plenum Press, New York, 3–12.

______. (1996a) *Eye-gaze avoidance and learning in orang-utans*. International Society for Comparative Psychology Conference, Concordia University, Montreal.

______. (1996b) *Gaze direction and visual attention in orang-utans*. XVI Congress of the International Primatological Society, Madison, WI.

______. (1998) *Eye-gaze and communication by organ-utans in zoos and in their natural environment*. XVIIth Congress of the International Primatological Society, Antananarivo, Madagascar.

______. (1999) Parental care in the common marmoset: anogenital licking and effect on exploration. *Journal of Comparative Psychology*.

Kaur, A. (1998) *Economic Change in East Malaysia*. Macmillan, London.

Kavanagh, M. (1983) *A Complete Guide to Monkeys, Apes and Other Primates*. Jonathan Cape, London.

Kawabe, M. (1970) A preliminary study of the wild siamang gibbon (*Hylobates syndactylus*) at Fraser's Hill, Malaysia. *Primates*, 11, 285–91.

Kehoe, M., Phin, C.S. and Chu, C.L. (1984) Tuberculosis in an orangutan pongo pygmaeus. *Australian Veternary Journal*, vol. 61, no. 4, p. 128.

King, B.J. (1994) Primate infants as skilled information gatherers. *Pre- and perinatal Psychology Journal*, 8, 287–307.

Kingsley, S. (1982) Causes of non-breeding and the development of the secondary sexual characteristics in the male orang utan: a hormonal study. In L.E.M. de Boer (ed.), *The Orang Utan. Its Biology and Conservation*. W. Junk Publishers, The Hague.

Köhler, W. (1921) *Intelligenzprüfung an Menschenaffen*. Springer Verlag, Berlin.

Lardeux-Gilloux, I. (1995) Rehabilitation centres: their struggle, their future. In R.D. Nader, B. Galdikas, L. Sheehan and N. Rosen (eds.), *The Neglected Ape*. Plenum Press, New York, 61–8.

Leighton, M. and Leighton, D.R. (1983) Vertebrate responses to fruiting seasonality within a Bornean rain forest. In S.L. Sutton, T.C. Whitmore and A.C. Chadwick (ed.), *Tropical Rain Forest: Ecology and Management*. Blackwell Scientific Publications, Oxford, 181–94.

Leighton, M., Seal, U.S., Soemarna, K., Adjisasmito, Wijaya, M., Setia, T.M., Shapiro, G., Perkins, L., Traylor-Holzer, K. and Tilson, R. (1995) Orangutan life history and vortex analysis. In R.D. Nadler, B.F.M. Galdikas, L.K. Sheeran and N. Rosen (eds.), *The Neglected Ape*. Plenum Press, New York, 97–107.

LeMay, M. and Geschwind, N. (1975) Hemispheric difference in the brains of great apes. *Brain, Behaviour and Evolution*, 11, 48–52.

Lethmate, J. (1982) Tool-using skills of orang-utans. *Journal of Human Evolution*, 11, 49–64.

Lewin, R. (1984) DNA reveals surprises in the human family tree. *Science*, 226, 1179–82.

______. (1988) Family relationships are a biological conundrum. *Science*, 242, 671.

Lewis, M. (1978) The infant and its caregiver: the role of contingency. *Allied Health and Behavioural Sciences*, 4, 469–92.

Lewis, M. and Coates, D.L. (1980) Mother-infant interaction and cognitive development in twelve-week-old infants. *Infant Behavior and Development*, 3, 95–105.

Lippert, W. (1974) Beobachtungen zum Schwangerschafts-und Geburtsverhalten beim Orangutan (*Pongo pygmaeus*) im Tierpark Berlin. *Folia Primatologica*, 21, 108–34.

Lovejoy, T.E., Bierregaard, R.O., Rankin, J.M. and Schubart, H.O.R. (1983) Ecological dynamics of tropical forest fragments. In S.L. Sutton, T.C. Whitmore and A.C. Chadwick (eds.), *Tropical Rain Forest: Ecology and Management*. Blackwell Scientific Publications, Oxford.

MacKinnon, J. (1971) The orang-utan in Sabah today. A study of a wild population in the Ulu Segama Reserve. *Orynx*, Journal of the Fauna Preservation Society, 11, 141–91.

______. (1974a) *In Search of the Red Ape*. William Collins, London.

______. (1974b) The behaviour and ecology of wild orang-utans (*Pongo pygmaus*). *Animal Behaviour*, 22, 3–74.

______. (1987) Conservation status of primates in Malesia, with special reference to Indonesia. *Primate Conservation*, 8, 175–83.

Maple, T.L. (1980) *Orangutan Behavior*. Van Nostrand Reinhold, New York.

Maple, T.L. and Hoff, M.P. (1982) *Gorilla Behaviour*. Van Nostrand Reinhold, New York.

Maple, T.L. and Zucker, E.L. (1978) Ethological studies of play behavior in captive apes. In E.O. Smith (ed.), *Social Play in Primates*. Academic Press, New York, 113–42.

Markham, R.J. (1990) Breeding orang-utans at Perth Zoo: twenty years of appropriate husbandry. *Zoo Biology*, 9, 171–82.

Marler, P. and Tenaza, R. (1977) Signaling behavior of apes with special reference to vocalization. In T.A. Sebeok (ed.), *How Animals Communicate*. Indiana University Press, Bloomington, IN.

Masson, J.M. and McCarthy, S. (1995) *When Elephants Weep. The Emotional Lives of Animals*. Delacorte Press, New York.

Masters, A.M. and Markham, R.J. (1991) Assessing reproductive status in orang-utans by using urinary estrone. *Zoo Biology*, 10(3), 197–207.

Matsuzawa, T. (1994) Field experiments on use of stone tools in the wild. In R.W. Wrangham, W.C. McGrew, F.B.M. de Waal and P.G. Heltne (eds.), *Chimpanzee Cultures*. Harvard University Press, Cambridge, MA, 351–70.

______. (1996) Chimpanzee intelligence in nature and in captivity: isomorphism of symbol use and tool use. In W.C. McGrew, L.F. Marchant and T. Nishida (eds.), *Great Ape Societies*. Cambridge University Press, Cambridge, 196–209.

McNeely, J.A. and Thorsell, J.W. (1989) Jungles, mountains and islands. How tourism can help conserve the natural environment. In L.J. D'Amore and J. Jafari (eds.), *Tourism: A Vital Force for Peace*. D'Amore and Associates, Montreal.

Meltzoff, A.N. (1996) The human infant as imitative generalist: a 20-year progress report on infant imitation with implications for comparative psychology. In C.M. Heyes and B.G. Galef Jr. (eds.), *Social Learning in Animals: The Roots of Culture*. Academic Press, New York, 347–70.

Miles, H.L. (1990) The cognitive foundations for reference in a signing orangutan. In S.T. Parker and K.R. Gibson (eds.), *"Language" and Intelligence in Monkeys and Apes: Comparative Developmental Perspectives*. Cambridge University Press, 511–39.

______. (1994) *Chantek: The Language Ability of an Enculturated Orang-utan.* International conference on orangutans: The Neglected Ape, Fullerton, California State University.

Mitani, J.C. (1985) Sexual selection and adult male orangutan pongopygmaeus long calls. *Animal Behaviour*, 33(1), 272–83.

Mitani, J.C., Grether, G.F., Rodman, P.S. and Priatna, D. (1991) Associations among wild orang-utans: sociality, passive aggressions or chance? *Animal Behaviour*, 42, 33–46.

Mohammad-Ali, K., Eladari, M-E. and Galibert, F. (1995) Gorilla and orangutan c-myc nucleotide sequences: inference on hominoid phylogeny. *Journal of Molecular Evolution*, 41, 262–76.

Moore, B.R. (1996) The evolution of imitative learning. In C.M. Heyes and B.G. Galef Jr. (eds.), *Social Learning in Animals: The Roots of Culture*. Academic Press, New York, 245–66.

Muir, C., Galdikas, B.M.F. and Beckenbach, A.T. (1995) Genetic variability in orang-utans. In R.D. Nadler, B.F.M. Galdikas, L.K. Sheeran and N. Rosen, *The Neglected Ape*. Plenum Press, New York, 267–72.

______. (1998) Is there sufficient evidence to elevate the orangutan of Borneo and Sumatra to separate species? *Journal of Molecular Evolution*, 46, 378–9.

Muir, C.C., Watkins, R.F. and Ha, T.T. (1998) Map simulations for testing a palaeo-migratory hypothesis. http://www.shef.ac.uk/%assem/4/4muwaha.-html

Munson, L. and Montali, R.J. (1990) Pathology and diseases of great apes at the National Zoological Park Washington, D.C. *Zoo Biology*, 9(2), 99–106.

Nadler, R.D. (1983) Experiential influences on infant abuse in gorillas and some other nonhuman primates. In M. Reite and N. Caine (eds.), *Child Abuse: The Nonhuman Primate Data.* Alan R. Liss, New York, 139–50.

______. (1988) Sexual and reproductive behavior. In H.J. Schwartz (ed.), *Orang-utan Biology*, Oxford University Press, New York.

______. (1995) Sexual behavior of orangutans. Basic and applied implications. In R.D. Nadler, B.F.M. Galdikas, L.K. Sheeran and N. Rosen (eds.), *The Neglected Ape*. Plenum Press, New York, 223–37.

Nadler, R.D., Galdikas, B.F.M., Sheeran, L.K. and Rosen, N. (eds.) (1995) *The Neglected Ape*. Plenum Press, New York.

Ng, F.S.P. (1983) Ecological principles of tropical lowland rainforest conservation. In S.L. Sutton, T.C. Whitmore and A.C. Chadwick (eds.), *Tropical*

Rain Forest: Ecology and Management. Blackwell Scientific Publications, Oxford.

Niemitz, C. and Kok, D. (1976) Observations on the vocalisation of a captive infant orang-utan (*Pongo pygmaeus*). *The Sarawak Museum Journal*, 24(45), 237–50.

Normile, D. (1998) Habitat seen playing larger role in shaping behavior. *Science*, 279, 1454–5.

The Northern Daily Leader (1997) They're weird: Monkey Business 1, Monkey Business 2. Armidale, NSW, 10.

Pagnel, M.D. and Harvey, P.H. (1993) Evolution of the juvenile period in mammals. In M.E. Pereira and L.A. Fairbanks (eds.), *Juvenile Primates. Life History. Development and Behaviour*. Oxford University Press, Oxford.

Parker, C.E. (1969) Responsiveness, manipulation, and implementation behavior in chimpanzees, gorillas, and orang-utans. *Proceedings of the Second International Congress of Primatology*, Karger, Basil, vol. 1, 160–6.

Parker, S.T. (1996) Apprenticeship in extractive foraging: the origins of imitation, teaching, and self-awareness in great apes. In A.E. Russon, K.A.Bard and S.T. Parker (eds.), *Reaching into Thought: The Minds of Great Apes*. Cambridge University Press, Cambridge.

Pearce, F. (1990) Hit and run in Sarawak. Under the floodlights of 24-hour forestry, the world's oldest rainforest is disappearing before the eyes of its inhabitants. Who will call a halt? *New Scientist*, 12 (May), 24–7.

Peltier, W.R. (1994) Ice age paleotopography. *Science*, 265, 195–201.

Penny, D., Murray-McIntosh, R.P. and Hendy, M.D. (1998) Estimating times of divergence with a change of rate: the orangutan/African ape divergence. *Molecular Biology and Evolution*, 15, 608–10.

Pereira, M.E. and Altmann, J. (1985) Development of social behaviour in free-living nonhuman primates. In E.S. Watts (ed.), *Nonhuman Primate Models for Human Growth and Development*. A. Liss Publishers, New York, 217–309.

Perkins, L.A. (1994) *International Studbook of the Orangutan*. Fulton County Zoo, Inc., Atlanta.

Perkins, L.A. and Maple, T.L. (1990) North American orangutan species survival plan: current status and progress in the 1980s. International Conference on Fertility in the Great Apes, Atlanta, 1989. *Zoo Biology*, 9(2), 135–40.

Peters, H. (1995) Orangutan reintroduction? *Department of Biology, Section of Ethology*. University of Groningen, Groningen, MSc thesis.

Phipps, M. (1992) Recent incidents involving illegal primate trade in Taiwan. *Asian Primates*, 2(2), 5–6.

Poole, T. (1987) Social behaviour of a group of orang-utans on an artificial island in Singapore Zoological Gardens. *Zoo Biology*, 6, 315–30.

Povinelli, D. and Cant, J.G. (1995) Arboreal clambering and the evolution of self-conception. *The Quarterly Review of Biology*, 70, 393–421.

Povinelli, D. and Preuss, T.M. (1995) Theory of mind: evolutionary history of a cognitive specialisation. *Trends in Neurosciences*, 18, 418–24.

Provine, R.R. (1996) Contagious yawning and laughter: significance for sensory feature detection, motor patterns generation, imitation, and the evolution of social behavior. In C.M. Heyes and B.G. Galef (eds.), *Social Learning in Animals: The Roots of Culture*. Academic Press, London.

Raven, P.H. (1988) Our diminishing tropical forests. In E.O. Wilson and F.M. Peter (eds.), *Biodiversity*. National Academy Press, Washington, DC, 119–22.

Redican, W.K. (1975) Facial expressions in nonhuman primates. In L.A.Rosenblum (ed.), *Primate Behavior. Developments in Field and Laboratory Research*. Academic Press, New York, 4: 103–94.

Reynolds, V. (1967) *The Apes: the Gorilla, Chimpanzee, Orangutan and GibbonCtheir History and their World*. E.P. Dutton and Co, New York.

Rijksen, H.D. (1978) *A Field Study of Sumatran Orangutans (Pongo pygmaeus abelli): Ecology, Behaviour and Conservation*. Veenman and Zonen, Wageningen, Netherlands.

Rijksen, H.D. and Rijksen-Graatsma, A.G. (1975) Orang-utan rescue work in North Sumatra. *Oryx*, 13, 65–73.

Rilling, J.K. and Insel, T.R. (1998) Evolution of the cerebellum in primates: differences in relative volume among monkeys, apes and humans. *Brain, Behavior and Evolution*, 52, 308–14.

Rodman, P.S. (1977) Feeding behaviour of orang-utans of the Kutai Nature Reserve, East Kalimantan. In T.H. Clutton-Brock (ed.), *Primate Ecology: Studies of Feeding and Ranging Behaviour in Lemurs, Monkeys and Apes*. Academic Press, London, 146–54.

Rodman, P.S. and Mitani, J.C. (1987) Orang-utans: sexual dimorphism in a solitary species. In B.B. Smuts, D.L. Cheney, R.M. Seyfarth, R.W. Wrangham and T.T. Struhsaker (eds.), *Primate Societies*. University of Chicago Press, Chicago, 146–54.

Rogers, L.J. (1997) *Minds of their Own. Thinking and Awareness in Animals*. Allen and Unwin, Sydney and Westview, Boulder, CO.

______. (1998) Orang-utans and other fauna. In A. Kaur and I. Metcalfe (eds.), *The Making of Malaysia*. Methuen, London, 47–68.

Rogers, L.J. and Kaplan, G. (1994) A new form of tool use by orang-utans in Sabah, East Malaysia. *Folia Primatologica*, 63, 50–2.

______. (1996) Hand preferences and other lateral biases in rehabilitated orang-utans, *Pongo pygmaeus pygmaeus*. *Animal Behaviour*, 51, 13–25.

______. (1998a) *Not Only Roars and Rituals. Communication in Animals*. Allen and Unwin, Sydney.

______. (1998b) Orangutans. In G. Greenberg and M.M. Haraway (eds.), *Comparative Psychology: A Handbook*. Garland Publishing, New York. 465–72.

Russon, A.E. and Galdikas, B.M.F. (1993) Imitation in free-ranging rehabilitant orangutans (*Pongo pygmaeus*). *Journal of Comparative Psychology*, 107(2), 147–61.

______. (1995a) Constraints on great apes' imitation: model and action selectivity in rehabilitant orangutan (*Pongo pygmaeus*) imitation. *Journal of Comparative Psychology*, 109(1), 5–17.

______. (1995b) Imitation and tool use in rehabilitant orangutans. In R.D. Nadler, B.F.M. Galdikas, L.K. Sheeran and N. Rosen (eds.), *The Neglected Ape*. Plenum Press, New York, 191–7.

Sayuthi, D., Karesh, W., McManahan, R. et al. (1991) *Medical Aspects in Orang-utan Reintroduction*. Proceedings on Conservation of the Great Apes in the New World Order of the Environment, Indonesia.

Schaller, G. (1961) The orang-utan in Sarawak. *Zoologica*, 46, 73–82.

______. (1965) Behavioral comparisons of the apes. In I. D. Vore (ed.), *Primate Behavior: Field Studies of Monkeys and Apes*. Holt, Rinehart and Winston, New York, 474–81.

Schlentrich, U.A. and Ng, D. (1994) Hotel development strategies in Southeast Asia: the battle for market dominance. In A. V. Seaton (ed.), *Tourism. The State of the Art*. John Wiley and Sons, Chichester, New York, Brisbane.

Schürman, C. and van Hooff, J. (1986) Reproductive strategies of the orangutan: new data and a reconsideration of the existing socio-sexual models. *International Journal of Primatology*, 7, 265–88.

Schwartz, J.H. (1984) The evolutionary relationship of man and orang-utans. *Nature*, 308, 501–5.

______. (1987) *The Red Ape: Orang-utans and Human Origins*. Houghton Mifflin, Boston.

Semendeferi, K., Damasio, H., Frank, R. and Van Hoesen, G.W. (1997) The evolution of the frontal lobes: a volumetric analysis based on three-dimensional reconstructions of magnetic resonance scans of human and ape brains. *Journal of Human Evolution*, 32, 375–88.

Shapiro, G.L. (1982) Sign acquisition in a home-reared/free-ranging orangutan: comparisons with other signing apes. *American Journal of Primatology*, 3, 121–9.

______. (1985) Factors influencing the variance in sign learning performance by four juvenile orangutans (*Pongo pygmaeus*). PhD thesis, Ann Arbor, Michigan, University Microfilms International.

Shapiro, G.L. and Galdikas, B.M.F. (1995) Attentiveness in orangutans within the sign learning context. In R.D. Nadler, B.F.M. Galdikas, L.K. Sheeran and N. Rosen (eds.), *The Neglected Ape*. Plenum Press, New York.

Shepherdson, D. (1994) The role of environmental enrichment in the captive breeding and reintroduction of endangered species. In P.J.S. Olney, G.M. Mace and A.T.C. Feistner (eds.), *Creative Conservation: The Interface between Captive and Wild Populations*. Chapman and Hall, London, 167–77.

Shively, C. and Mitchell, G. (1986) Perinatal behavior of anthropoid primates. In G. Mitchell and J. Erwin (eds.), *Behavior, Conservation, and Ecology*. Alan R. Liss, New York. 2, part A of Comparative Primate Biology: 245–94.

Smith, R.J. and Pilbeam, D.R. (1980) Evolution of the orang-utan. *Nature*, 284, 447–8.

Smits, W. (1998) Diary August 2–11, 1998. http://www.redcube.nl./bos/di980-804.htm (4 August), 1.

Smits, W.T.M., Herihanto, and Ramono, W.S. (1995) A new method for rehabilitation of orangutans in Indonesia. In R.D. Nader, B. Galdikas, L. Sheehan and N. Rosen (eds.), *The Neglected Ape*. Plenum Press, New York, 69–77.

Sodaro, C. A. (1988) A note on the labial swelling of a pregnant orangutan pongo-pygmaeus-abelii. *Zoo Biology*, 7(2), 173–6.

Spuhler, J.N. (1988) Evolution of mitochondrial DNA in monkeys, apes, and humans. *Yearbook of Physical Anthropology*, 31, 15–48.

Stanford, C.B., Wallis, J., Matama, H. and Goodall, J. (1993) Patterns of predation by chimpanzees on red colobus monkeys in Gombe National Park, 1982–1991. *American Journal of Physical Anthropology*, Suppl. 16, 186–7.

Sugardjito, J. and Nurhuda, N. (1981) Meat-eating behaviour in wild orang utans, *Pongo pygmaeus*. *Primates*, 22, 414–16.

Sugardjito, J., te Boekhorst, I. and van Hooff, J. (1987) Ecological constraints on the grouping of wild orangutans (*Pongo pygmaeus*) in the Gunung Leuser National Park, Sumatra. *International Journal of Primatology*, 8, 17–42.

Sugardjito, J. and van Hooff, J.A.R.A.M. (1986) Age-sex class differences in the positional behavior of the Sumatran orang-utan *Pongo-pygmaeus-abelii* in the Gunung Leuser National Park, Indonesia. *Folia Primatologica*, 47, 14–25.

Takasaki, H. and Hunt, K. (1987) Further medicinal plant consumption in wild chimpanzees? *African Studies Monographs*, 8, 125–8.

Tarling, N. (ed.) (1989) *Mrs Pryer in Sabah: Diaries and Papers from the Late 19th Century, introd. Nicholas Tarling*. Centre for Asian Studies, University of Auckland, New Zealand.

te Boekhorst, I.J.A., Schürmann, C.L. and Sugardjito, J. (1990) Residential status and seasonal movements of wild orang-utans in the Gunung Leuser Reserve (Sumatra, Indonesia). *Animal Behaviour*, 39, 1098–109.

Thompson-Handler, N., Malenky, R.K. and Badrian, N. (1984) *Sexual behaviour of Pan paniscus* under natural conditions in the Lomako Forest, Equateur, Zaire. In R.L. Shufman (ed.), *The Pygmy Chimpanzee*. Plenum Press, New York, 347–68.

Thorpe, W.H. (1963) *Learning and Instinct in Animals*. Methuen, London.

Tilson, R.L., Seal, U.S., Soemarna, K., Ramono, W., Sumardja, E., Poniran, S., van Schaik, C., Leighton, M., Rijksen, H. and Eudey, A. (1993) *Orang-utan Population and Habitat Viability Analysis Report*. IUCN/SSC Captive Breeding Specialist Group, Apple Valley, MN. 1–54.

Tobach, E., Skolnick, A., Klein, I. and Greenberg, G. (1997) Viewing of self and nonself images in a group of captive orangutans (*Pongo pygmaeus abelii*). *Perceptual and Motor Skills*, 84, 355–70.

Tomasello, M. (1993) Cultural transmission in the tool use and communicatory signaling of chimpanzees. In S.T. Parker and K.R. Gibson (eds.), "*Lan-*

guage" and Intelligence in Monkeys and Apes: Comparative Developmental Perspectives. Cambridge University Press, Cambridge, 274–311.

Tomasello, M., Call, J. and Gluckman, A. (1997) Comprehension of novel communication signs by apes and human children. *Child Development*, 68, 1067–80.

Tomasello, M., Gust, D. and Frost, G.T. (1989) A longitudinal investigation of gestural communication in young chimpanzees. *Primates*, 30(1), 35–50.

Ungar, P.S. (1995) Fruit preferences of four sympatric primate species at Ketambe, Northern Sumatra, Indonesia. *International Journal of Primatology*, 16, 221–45.

Utami, S.S. and van Hooff, J.A.R.A.M. (1997) Meat-eating by adult female Sumatran orangutans (*Pongo pygmaeus abelii*). *American Journal of Primatology*, 43, 159–65.

Utami, S.S., Wich, S.A., Sterck, E.H.M. and van Hooff, J.A.R.A.M. (1997) Food competition between wild orang-utans in large fig trees. *International Journal of Primatology*, 18, 909–27.

van Hooff, J.A.R.A.M. (1972) A comparative approach to the phylogeny of laughter and smiling. In R.A. Hinde (ed.), *Non-Verbal Communication*. Cambridge University Press, Cambridge, 209–41.

______. (1995) The orangutan: a social outsider. In R.D. Nadler, B.F.M. Galdikas, L.K. Sheeran and N. Rosen (eds.), *The Neglected Ape*. Plenum Press, New York, 153–62.

van Oosterzee, P. (1997) *Where Worlds Collide: The Wallace Line*. Reed Books Australia, Melbourne.

van Schaik, C.P. and Fox, E.A. (1996) Manufacture and use of tools in wild Sumatran orang-utans. *Naturwissenschaften*, 83, 186–8.

Vandevoort, C.A., Neville, L.E., Tollner, T.L. and Field, L.P. (1993) Noninvasive semen collection from an adult orangutan. *Zoo Biology*, 12(3), 257–65.

Veevers-Carter, W. (1991) *Riches of the Rain Forest. An Introduction to the Trees and Fruits of the Indonesian and Malaysian Rain Forests*. Oxford University Press, Singapore.

Visalberghi, E. and Fragaszy, D. (1990) Do monkeys ape? In S.T. Parker and K.R.Gibson (eds.), *"Language" and Intelligence in Monkeys and Apes*. Cambridge University Press, Cambridge, 247–73.

Visalberghi, E., Fragazy, D. and Savage-Rumbaugh, S. (1995) Performance in a tool-using task by common chimpanzees (*Pan troglodytes*), bonobos (*Pan paniscus*), an orang-utan (*Pongo pygmaeus*), and capuchin monkeys (*Cebus apella*). *Journal of Comparative Psychology*, 109, 52–60.

Visalberghi, E. and Limongelli, L. (1994) Lack of comprehension of cause-effect relations in tool-using capuchin monkeys (*Cebus apella*). *Journal of Comparative Psychology*, 108, 15–22.

Vochteloo, J.D., Timmermans, P.J.A., Duijghuisen, A.H. and Vossen, J.M.H. (1993) Effects of reducing the mother's radius of action on the development of mother-infant relationships in longtailed macaques. *Animal Behaviour*, 45, 603–12.

Voigt, J.-P. (1984) Behavioral observations within a group of young orangutans pongo-pygmaeus-abelii. *Zoologischer Anzeiger*, 213(3–4), 258–74.

von Koenigswald, G.H.R. (1982) Distribution and evolution of the orang-utan, *Pongo pygmaeus* (Hoppius). In L.E.M. de Boer (ed.), *The Orang Utan: Its Biology and Conservation*. Dr W. Junk Publishers, The Hague, 1–15.

Walker, E.P. (1964) *Mammals of the World*, vol. 1, The Johns Hopkins Press, Baltimore.

Weiss, M.L. (1987) Nuclei acid evidence bearing on hominoid relationships. *Yearbook of Physical Anthropology*, 30, 41–73.

Western, S. (1994) Ape Opera. *BBC Wildlife*, 10, 30–6.

Whiten, A. and Ham, R. (1992) On the nature and evolution of imitation in the animal kingdom: reappraisal of a century of research. In P.J.B. Slater, J.S. Rosenblatt and C. Beer (eds.), *Advances in the Study of Behaviour*. Academic Press, San Diego, CA, vol. 21, 239–83.

Whitmore, T.C. (ed.) (1987) *Biogeographical Evolution of the Malay Archipelago*. Oxford University Press, Oxford.

Whitmore, T.C. and Sayer, J.A. (eds.) (1992) *Tropical Deforestation and Species Extinction*. Chapman and Hall, London.

Wildlife Rehabilitation Statistics (1998). http://www.ndsu.nodak.edu/instruct/devold/twrid/ html/stats.htm

Williams, L. (1991) Born to be mild. *The Sydney Morning Herald Magazine*, 2 March, 34–5.

Winch, P. (1996) Orang-utangs have to be taught to climb. *The Sunday Mail*. Sydney, 16 June, 114.

Winkler, L.A. (1989) Morphology and relationships of the orangutan fatty cheek pads. *American Journal of Primatology*, 17(4), 305–20.

Wolpoff, M.H. (1989) Multiregional evolution: The fossil alternative to Eden. In P. Mellars and C. Stringer (eds.), *The Human Revolution: Behavioral and Biological Perspectives on the Origins of Modern Humans*. Princeton University Press, Princeton, 62–108.

Woodford, M.H. and Rossiter, P.B. (1994) Disease risks associated with wildlife translocation projects. In P.J.S. Olney, G.M. Mace and A.T.C. Feistner (eds.), *Creative Conservation: Interactive Management of Wild and Captive Animals*. Chapman and Hall, London, 178–200.

Xu, X. and Arnason, U. (1996) The mitochondrial DNA molecule of Sumatran orangutan and a molecular proposal for two (Bornean and Sumatran) species of orangutan. *Journal of Molecular Evolution*, 43, 431–7.

Yamagiwa, J. (1992) Functional analysis of social staring behavior in an all-male group of mountain gorillas. *Primates*, 33(4), 523–44.

Yerkes, R.M. and Yerkes, A.W. (1929) *The Great Apes. A Study of Anthropoid Life*. Yale University Press, New Haven, CT.

______. (1945) *The Great Apes*. Yale University Press, New Haven, CT.

Zhi, L., Karesh, W.B., Janczewski, D.N., Frazier-Taylor, H., Sajuthi, D., Gombek, F., Andau, M., Martenson, J.S. and O'Brien, S.J. (1996) Genomic differential among natural populations of orang-utan (*Pongo pygmaeus*). *Current Biology*, 6, 1326–36.

Zilles, K. and Rehkämper, G. (1988) The brain, with special reference to the telencephalon. In J.H. Schwartz (ed.), *Orang-utan Biology*. Oxford University Press, New York, 157–76.

Zucker, E.L. and Thibaut, S.C. (1995) Proximity, contact, and play interactions of zoo-living juvenile and adult orangutans, with focus on the adult male. In R.D. Nader, B. Galdikas, L. Sheehan and N. Rosen (eds.), *The Neglected Ape*. Plenum Press, New York, 239–50.

Zuckerman, S. (1981) *The Social Life of Monkeys and Apes* (reissue of the 1932 edition). Routledge and Kegan Paul, London.

INDEX